Praise for *The Persuasive Wizard: How Technical Experts Sell Their Ideas to Non-technical Decision Makers*

Lynwood Givens is a unique and gifted person. He is disarmingly smart. This is best demonstrated when he blends the talents of a highly-skilled engineer in a field that could truly be called rocket science with a very down-to-earth homespun approach to helping people appreciate the significance of the magic he is explaining. In my more than 35 years in the space business, there are very few people that can engage across the spectrum of people and create a special connection with all of them.

During our time together at Space Imaging, we endeavored to expose to the masses the power of remote sensing. Dr. Givens had the challenge of figuring out the myriad of applications that both industry and mere mortals would flock to. This work became the foundation for the geographic utilities like Google Earth that are now included in our everyday lives. Readers of this book stand to gain insight from a master.

Jeffrey K. Harris
Corporate Vice President, Lockheed Martin Corporation
Former President, Space Imaging LLC
Former Director, National Reconnaissance Office

(The National Reconnaissance Office controls all spy satellites and airborne reconnaissance for the US. The director reports to the Secretary of Defense. Space Imaging launched the world's first commercial imaging satellite. Lockheed Martin employs over 100,000 engineers.)

Some time ago, I had the pleasure of traveling throughout the Middle East with Lynwood Givens, watching as he made a complex American satellite system completely understandable to a wide range of political leaders, military officers and government bureaucrats. I marveled at this performance, as I had spent a couple of years as CIA's Director of Technical Services and experienced the rarity of this talent, even among the brightest of scientists and engineers. Now, Lynwood has written a readable and immediately useful guide to enable the rest of us to work this magic. This is a welcome and long-overdue book.

Frank Anderson
Former CIA Operations Officer, Central Intelligence Agency

(Mr. Anderson coordinated all CIA activities for the Near Middle East. He is frequently interviewed by major news sources and considered a national authority on the Middle East.)

One of the most difficult tasks facing the modern technologist is knowing how to translate creative ideas into concrete recommendations that are understood and accepted by the non-technical decision-maker. I experienced this dilemma first hand when, as Staff Director of the U.S. Senate Select Committee on Intelligence, I had to explain complex technical ideas to extremely intelligent, but non-technically trained, Senators. Dr. Givens has spent twenty-five years working with technologists and business leaders at all levels to bring these two worlds together. At Raytheon, I truly enjoyed our time working together. I came to know him as an expert communicator and a proven technologist, and I believe he is able to show others the techniques that work.

Taylor W. Lawrence, Ph.D.
President, Raytheon Missile Systems

(Prior to Raytheon, Dr. Lawrence was not only director for the US Senate Select Committee on Intelligence but also deputy director for DARPA, the Defense Advance Research Projects Agency.)

Dr. Lynwood Givens has a unique ability to distill complex technical subjects into simple terms easily understood by laymen or business executives. He is at ease communicating with senior executives or subject matter experts, such as scientists, mathematicians, or computer programmers. I had the good fortune to see this demonstrated time and time again, when Lynwood would present business proposals to captains of industry, leaders of government or internally at work. Often, Lynwood bridged a communication gap that not only spanned the technical vocabulary, but also the language as well, as we worked on numerous international proposals together. His advice and expertise in this area were of great help to me, and I am certain it will be to others, as well.

Barry M. Barlow
Department of Defense Presidential Rank Award Recipient

(The President of the United States bestows this award on only a very few select individuals within the Department of Defense.)

Dr. Lynwood Givens knows that, no matter how brilliant an idea, if the message won't reach and cause a reaction in its intended target, it has little value. He has worked with engineers who have not learned that a great concept must be sold in order to become reality, and has observed the critical skills that are needed to give life to an idea. His goal is simple, and that is to help others learn the ways that they can improve the chance that their great ideas can benefit others. This is a set of skills not taught in college.

Michael W. Mutek
Vice President and General Counsel, Raytheon Company

From growing up in the hardscrabble ranches and oilfields of West Texas, to starting an innovative medical imaging company, to convincing the nation's congressmen of the need for new multi-billion dollar a year aerospace programs, Dr. Lynwood Givens has led a very persuasive life. Lynwood is a gifted leader and spirited strategist. He evokes the epitome of thoughtful consensus building.

Having personally worked for and with Lynwood on large and complex problems of national and global importance in the areas of health, welfare and security, I can state that his persuasive methods work in moving large ideas and concepts towards realization. His personal life and professional business experience have honed his capabilities and expertise in this area to a fine art.

Timothy M. Stratman
Vice President, Kestrel Corporation

Dr. Givens' development of the steps in preparing a technical proposal for non-technical readers is an outgrowth of his over 40 years of experience as a researcher and administrator in the fields of solid state physics, optics, and space imaging. As one of Dr. Givens' former physics and chemistry professors and one who communicates with him routinely in his role as a member of the university's Board of Regents, I note the text carries Dr. Givens' characteristic approach: thorough, essential analysis and clear-thinking that made him a brilliant student, writer, and speaker whose ideas are readily grasped. His knowledge of the process from both ends — as a person or team member charged with writing a proposal and an administrator who must evaluate the merits of such proposals — serves Dr. Givens and his readers well, as he carefully lays out steps that lead to the kind of results that both parties seek.

Jesse Rogers, Ph.D.
President, Midwestern State University

In the complex world of Defense Electronics, success frequently requires "selling" technical ideas to non-technical decision makers — inside the company as well as the customer organization. A strong technical background is of little utility without the ability to communicate the ideas in a clear, concise and understandable fashion. I have had the pleasure of working with Lynwood Givens for many years and have been on the receiving end of a countless number of his presentations. He is extremely strong technically but his real skill is a unique ability to organize a diverse collection of complex interacting ideas into a free flowing technical presentation that can be understood by a layman. I can't think of anyone more qualified to write on the subject.

Terry Heil, Ph.D.
Senior Executive, Raytheon Company

Why I Wrote *The Persuasive Wizard*

"As president of this company, can you explain to me, please, why we have a dozen engineers and not one of them can communicate?"

"We wizards just want to do our jobs, but instead we are ripped from our computers, dressed up smartly, and dumped unprepared in a room full of executives we never heard of nor had any reason to know. We are expected to explain in minutes what required years to formulate, expected to use unfamiliar terms like "margins" and "pro forma" to explain concepts we understand only by algorithms and equations. And we blunt, factual wizards are told we must do this without stepping on any toes, rocking any boats, or perturbing the equilibrium of any wheeled vehicles toting apples."

Dedication

To my mother, Lillian Janice (Dougherty) Givens, who made this world a better place and taught her children they could be anything they desired – if they worked at it hard enough.

THE PERSUASIVE WIZARD

How Technical Experts Sell Their Ideas
to Non-technical Decision Makers

F. LYNWOOD GIVENS, PH.D.

Acknowledgements: I would wish to thank all who supported me in this work, but that is not possible. I cannot call by name the countless associates who taught me the ins and outs of technology, nor the managers and leaders who worked with me to put technology forward, but I am grateful to them all. To those who were kind enough to write reviews and recommend the book, I do give a special thanks and their names are included. I thank Bethany Beams for her diligent work in editing and proofreading the copies. I thank my sons Harrison and Nathaniel; Harrison, for his creative ideas, technology knowledge, photography, and for envisioning the name of the book; Nathaniel, for his continuing support and encouragement and willingness to listen to my thousand variations. To Janice, my wife of three decades, who kindly read portions, rendered phrases, and posed critical comments, thank you.

Email: *lgivens@thepersuasivewizard.com*

Visit my technology blog at
www.thepersuasivewizard.com
designed for technologists who need to persuade decision makers.

First Edition: 2011

ISBN: 1461198135

ISBN-13: 9781461198130

Table of Contents

Introduction

Stephen graduated from Stanford with a major in physics and telecommunications. In graduate school, he continued his research and developed unique technologies for signals transfer. His five patents caught the attention of executives at my Fortune 100 company. After several teleconferences, we flew to Salt Lake City hoping to capitalize on the inventions, close a deal and take the technology to market. Stephen's brother, the business major and CEO of their two-man company, expounded the benefits of their inventions, the ripe marketplace, and the lucrative investment opportunities. He was convincing. But, for an entire ninety minutes, Stephen sat across the table from us, fixating on his tennis shoes, fomenting differential this and integral that, fogging our heads with jargoned explanations, and drilling to the core of the planet with every protracted response. I believed Stephen had inventions of value, but I could put no words in his mouth, nor thoughts in his head, and we had only so much time—he failed to make his case with the president. When we left, I carried two things out of that office: my notes that would form a rejection letter to Stephen, and the sure knowledge that I would write *The Persuasive Wizard: How Technical Experts Sell Their Ideas to Non-Technical Decision Makers,* so Stephen, and thousands of technology wizards like him, need never fail again.

Good ideas are not lacking. Profitable schemes are not selling short. Technology opportunities are not dwindling. The paucity, the atrophy, lies in technical experts who can persuade non-technical decision makers to enact, empower, and invest. What is lacking is the wizard who can transition technology acumen into business action. *The Persuasive Wizard* teaches technologists everywhere, in all capacities, degreed or technician, how to persuade even top-tier decision makers for entrepreneurial funding, increased research and development dollars, or just promotion and advancement in their current jobs.

Chapter 1: The Art of Technology
Moving Past Science

The scientist only imposes two things, namely truth and sincerity, imposes them upon himself and upon other scientists.
— Erwin Schrodinger

Those wizards—the ones who create our technology, broaden our opportunities, and pave avenues for business—why are they such poor communicators? Why cannot they explain to us the methods and mechanisms of their work? Why cannot they describe, in words we understand, what is significant and what is not significant, when to research and when to manufacture, where to invest and where not to invest? Why cannot they show us how to help them, help us?

This is a worldwide phenomenon, that the titans of invention should be the pygmies of communication. The disease transcends age, gender, language, race, and locale. It stymies the high and the low. It plagues science, engineering, mathematics, computer science, telecommunications, medicine, networking, and the hundreds of other occupations where technologists ply their trade. In short, technologists are poor communicators.

We have a solution.

To understand why wizards cannot communicate with the outside world, examine how wizards are forged. They enroll in engineering, mathematics, computing, networking, science, and technology classes. They do not enroll in art. Those wizard courses are taught very differently than English, philosophy, and history courses.

What do they teach in those classes and how is it so different? They teach "rules," "structure," and "repeatability." They teach that every problem has *an* answer. They teach formulas to get those answers. They envision an objective world, not a subjective one. They instruct mathematicians, scientists, and engineers to find *the* one unique answer. In mathematics, the solution is called a "proof." It is not someone's idea, opinion, or viewpoint. This proof is unanimously endorsed by all other mathematicians as being *the* right answer, the only answer. Scientists are taught physical

$$f(x) = a_0 + \sum_{n=1}^{\infty} (a_n \cos \frac{n\pi x}{L} + b_n \sin \frac{n\pi x}{L})$$

$$\cos\alpha + \cos\beta = 2\cos\tfrac{1}{2}(\alpha + \beta)\cos\tfrac{1}{2}(\alpha - \beta)$$

$$(x + a)^n = \sum_{k=0}^{n} \binom{n}{k} x^k a^{n-k}$$

laws and the complex mathematical equations to describe those laws, exactly and precisely. Teachers convince engineers that every problem has *a* correct solution. Technologists everywhere work with applications and with data, massaging and testing until they get verifiable and repeatable results, the *same* results, at all times, in all places, under the same conditions, by everyone. Wizards always search for that one, correct, true solution. It is out there somewhere. It is not a matter of opinion or viewpoint; it is truth.

What are the elements of wizard education? Determinism, objectivity, and veracity. These modern-day specialists differentiate right from wrong, segregate truth from fiction, and quarantine correct from incorrect. Their problems may be complex, convolved, convoluted, and cantankerous. Their rigor may be abstruse, arcane, enigmatic, and paradoxical. Their answer may come through sudden epiphany or months of slave labor. But, rest assured, only one unique, correct answer will arise.

How does a trained technology wizard compare to the rest of the world? Contrast him to an artist, a sculptor, say. Let us consider this sculptor a genius, too, much like our wizard. How is this genius sculptor different from our technology wizard?

The sculptor walks through the marble quarry. He does not see tons of marble,

he sees what is inside the marble. He sees things others do not see. Michelangelo is the quintessential example of a sculptor. Consider his work *The Prisoners*. According to Michelangelo, there was never an "empty" block of marble, but always a prisoner inside. Now, that observation, itself, would be *like* our technology wizard. The wizard looks at technology problems and sees, or at least searches for, an answer. Michelangelo knows the prisoner is there, embedded in the stone yearning to escape. It is not Michelangelo's fault that others do not yet see the prisoner. Release him. Chisel him out. Let peers review, openly examine, and attest that the marble was never just marble but a prisoner. See, there is the prisoner. Look at him. He really was there all along. Michelangelo has chiseled him out. And, to the technologist, there is also the solution to be chiseled out. The wizard would look at every stone, that is, every problem, and likewise see a solution.

Yet, there are important differences betwixt the genius sculptor and our technology wizard.

First of all, Michelangelo chiseled four *Prisoners*, not one. View them, as I did, in Florence's Galleria dell' Academia. Not one of them is "finished," but they are considered masterworks of a genius. Five hundred years and those prisoners still struggle to escape the stone. I listened to my tour guide. I read his pamphlet. Not once did I hear or read the words "correct," "truth," or "repeatable." Michelangelo intimated that the prisoners were always there, embedded in the stone, and that he simply removed everything that was not a prisoner. Yet Michelangelo's marble and Michelangelo's chisel, in the hands of another—lesser or greater—would not have produced the same prisoners, and perhaps no prisoners at all. That is art. There is a result that earns approval, there is a result that merits disapproval, yet all judges will not concur. That is art. Rules do exist, but they are shadowy and ambiguous and are most often followed only by the novice. That is art, not science or technology.

Upon graduation, the savant engineer, scientist, mathematician, or technologist discovers that decision makers in the real world do not think the same way as wizards. The real world is different. It is more akin to art. In the real world, an outcome is seldom deterministic. Never is it certain. Always, it is difficult to control. Real world language is different. "Seat-of-the-pants" is a common term. "Who-you-know" is an important concept. "Extenuating circumstances" are frequently cited. "Correct," "true," and "exact" are rare forms.

LEVERAGING ART

Few disciplines successfully leverage technology with art, but architecture is one that does. An architect's design can follow all the rules: proportion, alignment, placement, and location. Stresses and strains can all be within specifications. Margins of error can be satisfied by every known calculation. Commercially available materials can be available and ready to ship. But, ugly is ugly and few investors build only that which is structurally sound, or "correct." When the bids are judged, art has a dominating influence on who receives the contract. Every competent architect knows how to specify the structural details, but what most often recommends them is creativity and art, not technology.

I toured the Robie House in Chicago's Hyde Park. Built over a century ago, it is a Frank Lloyd Wright masterpiece. On this hot, summer day, we started the tour outside. As a physicist, I instantly classified the construction as limestone, oak, brick, and glass. I speculated how one survived such a house in the age before air-conditioning.

Instinctively, I estimated the weight of the massive garage doors and calculated the torque on the hinge. I mentally caulked the leaks in this heterogeneous complex. I examined the joinery done individually and on site. I surmised a slump value (a factor used to measure the cement viscosity).

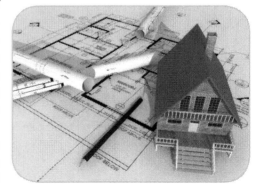

No one else on the tour seemed to have the slightest hint, inclination, or interest in such. What else was there?

I plied the tour guide for responses to my technology questions, but, she, like my wife who had reluctantly brought me along, seemed confused by this train of thought. Instead of technology, the architectural host opined the Prairie Style, the horizontal lines and the dramatic overhangs. She led a chorus of "oohs" at the art glass and "ahs" at the open floor plan. She never even mentioned the *slump*. (And, I know this house leaked like a sieve.) It was the art that dominated. In Ayn Rand's masterpiece, she foisted that her protagonist, creative architect Howard Roark, was even the *Fountainhead* of human progress. She was not talking about his technology. She was referring mainly to his art.

This melding of art and technology within architecture is an exception. Most technology disciplines never experience such a conjunction. Architecture is a model.

THE WIZARD APPROACH

We, senior executives at a Fortune 100 company, were introduced to Lauren when she entered the conference room for her first presentation to senior management. Hired straight from college, she spent the prior year working on a major proposal that we were now reviewing. Because of her promising abilities, the project manager selected her to present the software portion of our company bid. It was a scheduled 30-minute presentation for Lauren, but part of a longer eight-hour review for us. There was never any doubt in our minds that Lauren knew her software. The methodology was sound, obviously checked and validated many times by her. Nevertheless, from the beginning, it was obvious that Lauren had received every bit of our company's formal training on how to make a technology presentation—which is none. She had, however, tanked up on peer recommendations like, "Here's what worked for us last time," "Here's what we did at my former company," and "Here's the PowerPoint template everyone else is using."

Her presentation started late as the result of our overly scheduled president ending an overly long conference call. Lauren assumed that the entire schedule would shift and she would still have thirty minutes. Wrong. She began with an uninformative and vague introduction. She described her software with the mien of a scientist who had all day, meticulously annotating every point. Such details were undoubtedly crucial to her technology decisions, so she thought they must be important to us. When asked a question, she was defensive, almost offensive, taking time she did not have to expound on yet another detail that she thought we did not understand. Her time ended long before she did. She hastily threw up a useless summary chart and outwardly breathed a sigh of relief. So, inwardly, did we.

Afterwards, I queried the project manager about Lauren's assessment of her first presentation to senior management. The project manager relayed her reaction. "I was exhausted," she said. "My presentation was already longer than the thirty minutes allowed, and then they expected me to shorten it because *they* were late getting started. My analysis was complex. How could they ever expect to follow it with the chief financial officer dialoguing with the president during the most important part? How could they expect to understand my conclusions with incessant Blackberry poking, text messaging, and emailing? One of them looked at his email, left, and did not return. They were all so impatient they would not even wait for me to complete an answer to their questions. It takes a while to explain some of these things. No wonder I was not successful. They did not *understand* it."

Lauren encountered reality. She learned that decision makers and technologists think differently, but her conclusions and assumptions were in error. What she should have taken away was this: decision makers do not like equations and they find mathematics anathema. They refuse to learn jargon. They are uncomfortable with technology they do not grasp readily. They are impatient with technical details. They surmise that anything longer than a short answer is no answer. Their idea of the obvious differs from the wizard's idea of the obvious. They form unexpected conclusions based on external, non-technical, and sometimes tangential information. They want answers to their questions translated into a language they understand.

Lauren is a brilliant software engineer, but she never discovered an indispensable component to technology. She never discovered the *art*. There is an art of technology. That art is *persuasion*—getting across the value, the import, and the impact, communicating why it is important to the outside world, and what should be done about it. Some might call this *communication*, but I will explain now and later why *persuasion* is the right term. Ideas, even the greatest ones, will forever remain dormant, and atrophy, if they cannot be communicated adequately and supported for creation, enhancement, and implementation. Thus, if all a wizard does is *communicate*, then that will have little lasting impact. There are decisions to be made. If you just want to *communicate*, you may as well give decision makers a course in engineering and technology. What a wizard wants to do is not just give information, but simultaneously *persuade* decision makers that the wizard analysis is correct and should be supported, and that the wizard recommendations should be implemented.

There it is, then, the art to technology. This book is about the art of technology *persuasion* and how every technologist and scientist can learn that art. The technology wizards—hereinafter including all technologists in mathematics, science, engineering, computer science and programming, medicine, biology, geology—anyone whose job is technology, need never fear such a meeting as Lauren's if they learn the art. And it can be learned.

WHAT ABOUT YOU?

Let us see how this applies to you. Think about it. Your first presentation to senior management—if you are like most technologists, you are just glad when it is over. This "strange encounter" usually does not align with your personality and never was covered effectively in your education. Most wizards view a presentation to senior decision makers as a necessary evil, an agony best endured by others. Almost all wizards consider these meetings a waste of their valuable time, a dilution of resources, taking focus away from the real work, the work that actually matters—the research, the projects, the designs, and the inventions. The wizard sees these meetings as unnecessary consumption. After a typical encounter with senior management, most wizards, like Lauren, internally elucidate their feelings along the following:

You mean I went to school for this?

You've got to be kidding me?

Maybe I can beat it into them! [Meaning the decision makers]

They [the decision makers] *accepted another proposal over mine?*

It's not worth the effort!

After just a few of these frustrating presentations, the would-be purveyor of wizard technology prays never again to return to the conference room. And most find their prayers answered because the supervising gods would never, ever, again, put them in front of anyone in the decision-making process.

What happens if you seek guidance from seasoned veteran technologists? Technology peers and technology superiors mostly have the same response as you— presenting to senior management is a cross to be borne. And, unfortunately, the more brilliant the wizard the deeper this feeling of anxiety and the higher the hurdle. You ask peers and superiors, "What defines a 'good' technology presentation?" They reply unanimously, "One in which you survive. Count your body parts when you enter and count your body parts when you leave. Try to keep them the same."

WIZARD VS. WIZARD

Thomas Skidmore earned his Ph.D. in mathematics from an Ivy League school. He worked in important organizations throughout his career, excelling in most of them. He was an acknowledged expert in his particular field, a field I cannot openly explain because it involved a top secret US intelligence organization. We were in heavy competition for a major contract. The vice president of our division, knowing Skidmore mostly by reputation, demanded Skidmore make the technology presentation at the meeting with top-tier senior government officials. The vice president reasoned that the best technologist would give the best technology presentation, which would give us the best technology bid, which would beat our competition in technology. What is wrong with that?

I had known Skidmore for several years. No one would deny that he was the best among us, a brilliant mathematician and a creative thinker. If you were working on a problem and ran out of ideas, you went to Skidmore. He would listen to the arguments, think a moment, and then come up with three or four new things for you to try. Good things. Things that worked. Sometimes, he could just look at your analyses that had taken you days and, in just a few minutes, see the error of your thinking and show you how to correct it. He was brilliant.

He was also erratic, a Boy Scout in suspended adolescence.

Back to the meeting with the heads of intelligence and our need to win the contract. Skidmore was a capable speaker among *peers*. He usually spoke *ad lib*, but since this was an important meeting, he wrote his notes on 3"x5" cards, so he could get everything *just right*. The cards were unnumbered and mostly incoherent to anyone else: scribbles on this one, underlines on that one, front and back, and so forth. But they undoubtedly worked for him. (Of course, they were in pencil.)

He is seated at the twelve-person conference table beside Dr. Charles Lebreau, a consultant of ours who is in the same intellectual league as Skidmore. We brought in the renowned Lebreau for just this meeting. In addition to being brilliant, Lebreau has his own quirks, those of being curious to a fault and easily distracted.

I look across the table at these two geniuses sitting side by side. Skidmore is scheduled to be next with his presentation. He has his note cards stacked neatly in front of him, ready to go. He is looking at the current speaker, listening attentively, looking ahead to a quick finish to his presentation so he can go back to the laboratory to what he calls "the real work." Lebreau is in his own world, which is nowhere near this conference room. I see Lebreau reach over and take the bundle of 20-30 note cards that Skidmore plans to use. I think, "Oh, my ..."

Lebreau starts reading them, laying the cards out in front of him. He starts playing his own version of solitaire, not a care in the world, glancing at this card, shuffling that one, upside down, sideways, spinning them. Shuffle, shuffle, shuffle. Spindle, cut, shuffle.

The room lights come up. They announce Skidmore's turn to present. Lebreau wakes up from his card game, shuffles the deck, and pushes them back under Skidmore, who is looking elsewhere. Skidmore obliviously grabs the disorganized cards and marches to the podium with the confidence of Napoleon. He starts the presentation with the first card. Then the second. Another. After a few minutes, even he realizes that his notes are totally out of order with any logic he may have had and also totally out of synchronization with the computer graphics he is clicking on the overhead. He has no idea how to put the notes back in order. A lesser mortal would have numbered them, but not Skidmore. Unable to disentangle the Gordian knot, he simply continues speaking one card at a time. The summary is somewhere in the middle, the introduction near the end, the data are spread throughout, and, well, you get the picture. He knows they are out of order, of course, but his stream-of-consciousness reconstruction does nothing but foment an incessant reshuffle, going back and forth, putting this one down, picking that one up, turning this one over, musing on that one, a "hmmm" here, a "never mind" there. The projector continues in its fixed, now unrelated, sequence. It is the "perfect storm" of a presentation.

During all this, of course, the unabashed Skidmore continues to think himself brilliant. He is oblivious to the chagrined and nonplused teammates such as myself. The government decision makers are scratching their heads and chuckling out loud, probably trying to figure out what the joke might be. Our technology solution was complex and now impenetrably so by the convoluted, erratic, and disjointed presentation of Skidmore. Skidmore finally ends our misery, a full hour past schedule. His closing statement is a hackneyed, "I know this has been a little difficult to follow, but it's still probably good enough for government work." It was not.

We were not awarded the contract, a conclusion any fool at the meeting could, and did, make. Any sense of personal responsibility for the loss of this contract was transparent to either Skidmore or Lebreau, who continued to be Skidmore and Lebreau. Our vice president never demanded another presentation from the renowned Dr. Skidmore, and neither did I when I became his boss.

I wish I could say that this story was fictional, anecdotal, or isolated, but it is none of those. It is rerun every day by those technology wizards who have never learned the art of technology persuasion.

THERE IS A CHASM

The world of the wizard and the world of the decision maker are vastly different. The Jewish patriarch, Abraham, yelled from Heaven across to where Dives resided in Hell. "Between us and you a great chasm has been fixed, so that those who want to go from here to you cannot, nor can anyone cross over from there to us." The world of the wizard and the world of the decision maker may not be as different as that of Abraham and Dives, but they are different. Some would envision, even, that heaven and hell are apropos. All have seen the chasm.

Why is there such a void between the technology wizard and the decision maker? Why are the two domains so different? Because the womb that conceives an idea may be ill-equipped to birth it. There is a vast chasm between conception and completion, between formulation and formation, and between innovation and implementation. A persuasive wizard must bridge this chasm that separates her innovative, creative genius from the practical, realistic, nuts-and-bolts world of implementation, which is hard to do. With the right training, a technologist can bridge this chasm and make recommendations, time and again, that are successful, effective, and, more importantly, approved.

It is a difficult bridge, this technology persuasion. The typical working wizard is not a nationally recognized figure, which would make it easier. If a wizard were nationally recognized, she could just say things and, though no one would understand them, they would think the things must be true because she is a nationally recognized wizard. No, most of us are thought of as experts in our own little narrow domain, but few are known outside their restricted areas. And most of us are perfectly content to keep it that way. We delight in Dilbert cartoons and think they represent management to a tee. We have no problem being labeled as geeks or nerds. When we are asked to present to the CEO, the president, the general manager, or just our boss, we feel put out, awkward, and out of sync with the request.

We technologists just want to do our jobs but instead we are ripped from our computers, dressed up smartly, and dumped unprepared into a room full of executives whom we never heard of nor had any desire to know, anyway. Our presentation is wedged into an inadequate time slot where we are expected to explain in minutes what took months and years to formulate. And we blunt, factual wizards are told we must explain our technology without stepping on any toes, rocking any boats, or perturbing the equilibrium of any wheeled vehicles toting apples. We are plunked from the research facility, the software group, the hardware group, from systems, from the medical team, the chemistry table, the physics calculations—from drawings, equations, graphs, models, simulations, tables, calculations, computers—and plopped down in the domain of the titans. There, we are confronted by behemoths seeking nothing but immediate results, emperors challenging us for returns, and monarchs warring about margins and other things we know nothing about. We hear about investors, stockholders, and boards (lions and tigers and bears) nipping at the heels of our very livelihood. It is not easy to make effective presentations when you really do not know what to expect or what is expected of you.

This chasm between wizard and decision maker is vast. What we desire to do is to persuade the decision makers to make their decision in our favor, one that advances our love for technology, for research, for projects, and for our innovations. We see our love for technology as being also in the best interests of the decision makers. It is not just for our good, it is for their good. It is for the good of the *world*. We technologists live on high moral ground. We seek to persuade others of the *truth*.

Recall that I specifically use the term technology *persuasion*, as opposed to technology *communication* for a reason. But, by persuasion, I do not mean to imply in any way that a presentation should "spin" the audience of decision makers to support inadequate or marginal ideas. Perhaps one can do that, and possibly it is done routinely, but that is not any part of this book.

In the 5th century BC, Athenian democracy permitted any adult male the opportunity to speak at the weekly *ekklesia*, a public gathering and the source of Greek political policy. Since persuasion was a key factor in speaking at the *ekklesia*, a group of teachers formed whose purpose it was to teach young men the skill and art of persuasion. Their technique was not to use force of logic, but to find words, rhetorical techniques, and clever phrases that would persuade and move an audience, notwithstanding the merit of the facts otherwise. These teachers were the Sophists. Hence the word "sophism" has come down to us as a person who uses specious arguments to persuade, often for the very intent of deception. Sophism is not a component of wizard persuasion.

The purpose here is to bridge the gap between the wizard and the decision maker so that good ideas will be accepted and given a chance to take root. The purpose here is to understand how to present technology concepts exactly, but effectively.

I limit here the use of the term communication because "communication" has too broad a connotation and too weak a definition. One does communicate technology, *per se*, to peers, using the language of the particular technology discipline. But decision makers are mostly non-technical people and not peers. They usually do not want technology communicated to them. For what purpose would one *communicate* technology to a non-technical person? Are non-technical decision makers suddenly fascinated with differential equations? Have they read a book on SCSI and SATA and cannot contain themselves? Did they pick up their *electronic-this* and *internet-that* and suddenly yearn to solve Maxwell's equations? Probably not.

Non-technical persons ask about technology because they want an answer to a question that affects *their* lives, not the life of the technologist. They want to make a decision about which network to install, or which surfactant to deploy, or which innovation will reap the greatest margin. They want to understand technology at an investor's level, how it impacts them. In my training sessions with wizards, I often stop them while they are still in their introduction. I say, "I'm a decision maker. Tell me why I should care about what you just said. I understand why it is important to you. Your job is to tell me why it is important to me."

CHASMS BROAD AND DEEP

Now, given any audience and any subject, what the speaker intends to say and what he actually says are not necessarily what the audience hears. This is especially true when the presentation is of a complex technical nature and the audience is non-technical. Most wizards cannot overcome this "impedance mismatch" (an electrical engineering term).

It is not unusual for a technical presentation to actually destroy the underpinnings that otherwise would have supported it. Skidmore is a case in point. Instead of moving decision makers to make additional investments, a poor technology presentation can actually terminate the funding. This occurs not because the decision makers are enlightened and terminate based on prudent business decisions, but because the technical presentation is so tangential and confusing that it leaves their minds (followed closely by their pocketbooks) free to champion other causes they do understand.

Really brilliant wizards can get a project killed posthaste.

Throughout my career, I have had outstanding technologists on my team, world-class geniuses—engineers and scientists—revered by their peers. They were published, received awards, and were broadly recognized for their technical and scientific acumen. But they could be deadly when in front of management. They were nothing short of suicidal terrorists when they talked to our customers. They could nuke a project before they finished the introduction. If I was ever instructed by senior leaders to have these people present to management, I made every effort to have damage-control and recovery teams on full-alert standby. It was obvious to me why we always had these geniuses working in closed, secured areas. Outsiders think the purpose of the security was to keep the spies out. It is not. It is to keep those wizards in.

Such need not be the case. This persuasion ineptness is both innate and learned, but not impossible to change. A different outcome can and will result if the technologist, the wizard, is willing to learn the art of technology persuasion.

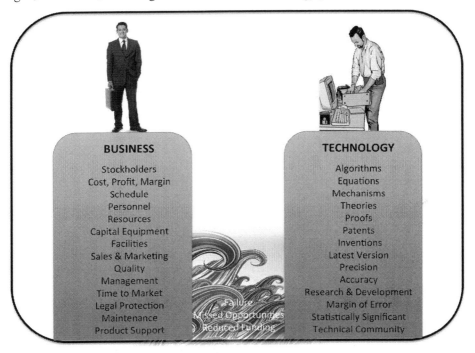

Figure 1-1: Business and Technology Are Different

BRIDGES THAT SPAN THE CHASM

Isaac Newton, that giant among us, said, "We build too many walls and not enough bridges." Apparently, he knew this well because all his biographers document his conflicts and frustrations even with peers. While the wizard inclination is, indeed, to build a high, impenetrable wall, the successful technologist leans how to turn wall building into bridge building.

A Colorado legal firm met with me. They wanted to show that effluents from a local manufacturing facility were migrating into a nearby stream and adversely impacting their client's trout-fishing business. They did not know what technology was available, but they did know what they needed from a lawyer's standpoint. They did not describe it in these terms but they needed non-invasive evidence, measurements, and identification of effluent migration. They wanted to know *if* we could do this, *when* we could do it, *what* it would cost, how *accurate* was it, and how could it be defended in court. Did they want to discuss the near-field implications of ground-penetrating radar? No. Did they want to understand solutions to the Navier-Stokes equation? No. They wanted technology persuasion.

I met with the US Department of Agriculture and said, "When it hails in Abilene and destroys the wheat crop, each farmer assesses his damage and files a claim. How do you, the USDA, verify the extent of the damage and field ownership?" Their answer was, "We hire someone to drive around in a pickup truck, look at the fields, and make an estimate."

We were in the Northwest discussing pipeline transportation. I queried, "Over all this rugged terrain, how do you monitor the pipelines for leakage and fatigue?" Their answer was, "We hire a pilot. He flies along the pipe. He looks out the window. He files a report."

I posed the question, "If there were a better way, a cheaper way, a more accurate and timely way to do your job, would you change what you are doing?" "Of course," they replied. And, at that point, the technology persuasion began. We had technology, yes, but they had to be convinced first, that it would do the job they wanted to do. They had to appreciate the value proposition. They had to understand reliability, accuracy, speed, and implementation. It was about technology persuasion.

The wizard must learn to persuade in such a way that even though the audience does not understand the technology, they understand how it impacts them. Not in a car-salesman (forgive me) or infomercial way, but to understand the technology sufficiently at their level to make an honest, informed assessment. Not an assessment through emotion or embarrassed ignorance, but a genuine movement from not knowing to knowing.

To move a non-technical person from analysis to acceptance is not a deterministic

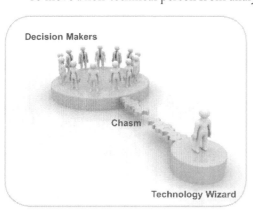

process. That is why we started with the term *art*. Can a technologist learn how to deliver recommendations that will be followed by non-technical decision makers? Yes. It is an art that, with practice, can be learned. Any wizard of technology can learn to effectively persuade even the most senior of decision makers. Any technologist can learn to communicate complex technical information in an understandable and effective way.

CHAMPIONS THAT CROSS THE BRIDGE

When President John Kennedy announced to the world that the Soviet Union had put missiles in Cuba, it was not Secretary of Defense Robert McNamara who persuaded him. McNamara, himself, was working the political agenda. It was a *technologist*, a CIA wizard at the National Photographic Interpretation Center (NPIC), who persuaded Kennedy that the little fuzzy blobs in the U-2 photographs really were Soviet missiles. Kennedy knew nothing about aerial photography. To the novice, aerial photographs are impossible to unambiguously interpret. If you have never seen a Soviet missile from overhead, you will not recognize it in a photograph. Kennedy had not and did not. Those images, now unclassified, were only about 95% confirmation to an expert and zero percent to a novice. Kennedy was a novice. Kennedy had the images right in front of him, there on the table, but he had not the wherewithal, had not the technical forte, to say with any knowledge that they were missiles, Soviet or otherwise. McNamara was no better. The *technologist* persuaded the President that they were missiles. The technologist knew his business, but he also knew the art. It was not just the pictures; it was persuasion—technology persuasion— communicating the value proposition.

A few years ago, Ken was the manager of one of our research groups. He realized the impact of inserting advertising banners in front of the baseball bleachers during a broadcast, or drawing down-markers on the football yard lines in a football broadcast—banners that never exist and markers that are never there—all a trick of the then-emerging digital graphics, now used routinely in every major sports broadcast. When Ken first had this idea, he left his six-figure, secure job, put together a business plan, and sought investors. Ken knew the art. I met Ken at a celebration with his employees. They had achieved yet another profit milestone. Ken persuaded the investors because he learned the art.

Eleanor went to school part-time while being paid full-time to work for our company. How? She made her case before management, the company agreed with her potential, paid her to go to graduate school while still "employed," and a few years later we celebrated her doctoral degree. Recently, she finished the software design for a billion-dollar system for our company. She learned the art.

Phil had an idea for telecommunications band-hopping and developed it into a product. Phil is not a skilled presenter, not by any means. In fact, he is the stereotype of the very person targeted by this book. He is a genius, a true wizard, but he cannot speak in your world. He can spit ATM's, BPSK's, CDMA's, IPVn's, IGFETS, and MOSFETS until your little glassy eyeballs pop out and roll across the floor. He never grasps that you dropped synchronization hours ago and all the circuits are jammed. He will just keep going, and going, and going. I worked with Phil to help him formulate things in a different way because persuasion is a skill that can be learned. The last time I spoke with Phil he had obtained funding for a new company, was working on a prototype and hoped to close shortly on a major telecommunications contract. He is learning the art.

What power do all these technical experts and others like them share? They know their technology and they possess the power to persuade non-technical decision makers.

WHY THIS BOOK WILL MAKE YOU SUCCEED

This book targets the innovative creative genius who lacks these communicative skills and this persuasive knowledge. It transforms the techno-babbling wanna-be into a persuasive force for entrepreneurial funding, for research investments, for increased staffing, for new product development, and for the hundreds of other dreams that wizards have.

How does one develop this persuasive power? The traditional speech-making guidelines and recommendations are often inappropriate and always ineffective for technical experts persuading non-technical decision makers. The speed of business, the flux of communication, the complexity of technology, the guerilla business strategies, the exponential velocities of technology development: all these have produced a new environment. It is an environment of turbocharged decisions, induced attention-deficit disorder, an ear for sound bites, an eye for video clips, and a mind for fragmented multiplexing.

Whether you are a new graduate or a veteran technologist, this book describes every step of the process for you, including idea creation and persuasion strategy. It answers questions like how to originate technology concepts, where and how to start the strategy, when to advance and when to back off, when and how to end, what to do in between, what to emphasize, what to leave out, what to wear, what to expect, how to answer questions and indeed, the complete repertoire.

Do technical experts feel their weakness as communicators? You bet they do. Do they lose sleep worrying over it? You better believe it. Are they genuinely affected by those shortcomings so obvious to outsiders, internalized to themselves, and stereotyped by observers? Make no mistake—that external wizard aloofness, that seeming indifference, is fragile veneer.

When managers and presidents and CEO's formulate new markets, take on new partners, and claw for growth of their companies, do they desperately need their technologists? They do. At that most critical moment in the negotiations when they must close on a technology, do they pray that their technical experts will communicate effectively to sway internal factions and external investors? Absolutely. Will those executives do everything in their power to bring their competitive technology to the forefront? Yes, time and again. The reason CEO's put words in the mouths of their technologists is because the technologists are unable to put words there, themselves. The instructions in this book will change that.

Understanding this art of technology is not easy. Most books on communication and public speaking treat technology presentations as "how-to" speeches: how to do this or how to do that. Or, if not, they lump them into the standard communications presentation. There are worlds of differences in how each combination of speaker and audience should be approached. A technical presentation to decision makers is a heterogeneous interaction requiring precise technical knowledge, general business knowledge, circumspection, and perspicacious communication skills. It is a crossing of two very different worlds. It is a rare technologist that can communicate effectively to any audience, much less an audience of diverse top-tier decision makers.

When I first started leading technology groups, I read books on how to communicate and took courses in how to make effective presentations. I was taught what to say and what not to say. I read books on how to influence and articles on how to convince. I attended lectures on how to communicate. I took professional training. I hired tutors. I listened to feedback from peers, subordinates, and supervisors. I constantly strove to be an effective purveyor of technology ideas. Then the epiphany occurred. The books and courses were misdirected, even wrong.

Those books and courses were not designed for technology persuasion; they were not designed to teach one how to connect the world of the technologist to the world of the decision maker. They gave instruction and guidance in general communication and organizational skills, true, but there was scant inclusion of technology. More and more I realized that many of the accepted presentation guidelines were noticeably ineffective when applied to the particular situation of technologists communicating with decision makers. Something more, much more, was needed.

In practice, I discovered just how impotent were some of the current guidelines and techniques. Oh, did I. I hit upon why this was so. First, most authors writing in this field are professional *communicators*. They are not degreed scientists or technology experts. They have little or no formal technology experience and even less knowledge of the people who do. Technology wizards have special needs, but the writers do not know what those needs are or how to compensate for them.

Second, the authors, although possibly renowned as motivational speakers, have little relevant experience with heads of corporations making technology business investments, legal attorneys deciding technical license and patent rights, heads of government making technical purchases, high-ranking military officers investigating new weapons of war, intelligence organizations evaluating major new capabilities, or anything similar to this. To have technical credentials, experience with senior-level business executives in these organizations, and the ability to write about them is close to unique. Here, we attempt to do just that.

There are other reasons why past methods fall short and why it is time for new ideas. Times change and people change. The decision maker's world is a cacophony of information overload, telephones, text messaging, constant interruptions, meetings, and handheld interfaces. Any presentation to decision makers must be in syncopation and sympathy with these other perturbations.

Successful technology persuasion today requires new, innovative ways of thinking and presenting. It is totally different than it was even ten years ago. It is a different paradigm. Most decision makers today do not come up from among the technology ranks. Most decision makers today know far less about technology than did their predecessors; their lifelong exposure is becoming more about the use of technology and less about the mechanisms of technology.

Many of the ideas in this book run counter to what is taught in traditional communications courses. That is so because this is not traditional communication. This is persuasion of the most complex technology ideas to the highest level of non-technical individuals, who will make decisions based on the information you present them. The standard communications paradigm is not designed for this complex conjunction of disciplines.

To further refine what works and what innovations are needed, I worked with high-tech businesses and governments around the globe, making thousands of presentations to company officers, government officials, senior military commanders—decision makers all—around the world, presentations to presidents, chief executive officers, chief operating officers, chief financial officers, chief technology officers, directors, boards, legal departments, human resources departments, contracting departments, heads of government, teachers, professors, associates and competitors, all of these and more. During this time, I examined the outcome of these presentations to refine the techniques and methods that ensured a high percentage of success for even the most complex technology ideas and for even the most technically naive audience.

Not every decision maker agrees on every detail of this book, nor is it for every situation. That is what makes it an art. But it is a *learnable* art whose precision is improved with the correct instruction. Exercise these guidelines and the outcome will be a technical presentation that motivates decision makers to enact major technical programs and initiate major investments on your behalf.

During my career, I directed teams of engineers and scientists, men and women, who created and invented products of which others only dream. They fashioned weapons of war and invented instruments of peace. They created intelligence engines, patented inventions, and revolutionized communications. They were wizards, many of them world-class. As chief technology officer, I reviewed idea after idea for funding.

Many of those ideas were rejected by me and others. Had those rejected ones been presented correctly, management would have funded many of them and consequentially harvested significant returns for the company. Instead, the companies endorsed other ideas (those presented correctly to management), while the poorly presented ones foundered. Competitors picked up some of those discarded concepts and ran with them. Admittedly, a few of those technology ideas were properly jettisoned and some, maybe, should have been torpedoed sooner. But I speak here of the majority, those ideas that had legitimate and solid technology and business merit, but failed to make their case in the conference room. And, those were a loss for all of us: inventor, investor, and consumer alike.

When persuasive skills are developed fully, when this art of technology persuasion is mastered, it has two marvelous effects. It drives the wizard to vastly improve not just the technology persuasion, but the technology *itself*. It actually improves the technology, in ways that I shall show later, and in doing so, propels the decision maker to endorse those wonderful ideas that were so wonderfully presented.

Moving Past Science

- Technology persuasion is an art.
- Mastering the art improves the technology itself and wins approval in the board room.
- Persuasion goes vastly beyond communication of technology. It melds technology and business.
- This art is not simply a description of technology at the layman's level.
- It is not "spin." It is effective persuasion of essential elements crucial to the decision.
- It is the transferal of technical ideas to a non-technical decision maker in ways that foster a favorable decision for the technology and the business.

Chapter 2: Strange Encounters
Understanding the Non-technical
Decision Maker

We must do that which we think we cannot.
— Eleanor Roosevelt, First Lady of the United States

*T*he distance from the chief executive officer to the podium is thirty-two feet. It looks much farther from the podium end. Twelve board members are seated around three sides of the massive rectangular table that defines the room. The CEO leans back at one end; the speaker leans forward at the other. The speaker of the moment is a research scientist, always a risky move. His supervisor is in the back of the room with the other layers of management. The scientist describes a software application designed and built by his group and alpha-tested by the software organization. The company wants to take this application public. It is a tough sell. This is a mostly systems-oriented electronic hardware company. A commercial software package is out of their comfort zone. This is the reason the board must first approve it before it can go forward.

The board asks questions. The president asks questions. They query the vice presidents of the various departments: engineering, finance, marketing, sales, and legal. The interlocutors do not have all the answers, but they do have good answers and no one sinks the ship. The scientist is requesting approval and funding to move forward with beta-testing.

The board discusses in muted volume. The CEO leans back further. The speaker stands erect. The one leaning back makes his decision.

The product is headed for market.

Technologists make recommendations. Decision makers make decisions. Those decisions enable, reject, or stalemate the recommendations.

What constitutes a decision maker? The decision maker is that person whose position is so high, or influence so great, that others defer. At the end of your technology presentation, if all eyes pivot to a single individual, that person is the decision maker. Sometimes the decision maker sits at the head of the table, but not always. The person at the head of the table usually has the highest title, but may not be the decision maker for your specific requests.

Others may influence the decision maker's decision; others may formalize the decision maker's decision. Still others may articulate the decision maker's decision. But, in the end, the decision maker makes the decision. Now, these other individuals are not *the* decision maker, true, but ignore them at your own peril. Like chess, the king will decide the game, yet focus all your attention on the king and you invite failure.

Whether you are meeting with your immediate supervisor or presenting to the chief executive officer, someone in the room has the authority, or will take the responsibility, to make the decision. There may be the appearance of a vote. There may be the semblance of democracy. This is usually form and not substance. By definition, the decision maker decides.

To understand decision makers, we need to examine them. It is impossible to persuade if we do not know who we are persuading, or what will persuade them. Are decision makers all alike, for instance?

Decision makers have similarities, just like wizards have similarities. In some respects, I find decision makers to be as different from each other as wizards are different from each other. Among decision makers, the backgrounds, thought processes, influences, acumen, patience, perspectives, integrity, insights, and knowledge can be different, but the approach to making decisions has some commonality. Studies in executive decision making and studies in decision theory reveal that there are characteristics common to decision makers. Knowing these characteristics will help you understand and relate to your own individual decision maker.

CHARACTERISTICS COMMON TO DECISION MAKERS

What traits are common to decision makers? How are they characterized? How are they alike? In surveying decision maker characteristics, the following descriptions appear repeatedly.

Ambitious—executives in decision-making positions want to make a name for themselves: the bigger, the better, the sooner, the better. They want to achieve something worthwhile and noteworthy, maybe even notoriety. Subordinates do not accuse them of thinking small. Decision makers have lofty goals—personal, if not corporate. Ambition may derive from selfish motives or innate passions, usually from a combination of the two. Decision makers push themselves, and others, to levels ordinary people consider uncomfortable and extreme. That which decision makers see as reasonable and achievable may be viewed as Herculean by subordinates. Ambition can drive some decision makers to constantly be concerned with details, while it can push others to surf upon superficial, broad generalizations. Ambition can ensure solid, well-thought-out decisions or, conversely, can produce spontaneous decisions lacking substantiation. In extreme cases, ambition can drive a decision maker to corruption and total disregard for the organization, family, and person.

Driven—they operate in a single mode, all-out, the accelerator floor-boarded, the engine red-lined, and every tire smoking. What subordinates see as requiring days to do is expected in hours. Months are days, years are months, tomorrow is now, on time is late. They work 24/7. Whatever they do, whenever they do it, wherever it is done, they are on the job. Daytime, nighttime, weekends, eating, drinking, traveling, sleeping, dreaming: they are all the same to them because decision makers are not elsewhere, they are on the job. They drive no one as hard as they drive themselves. Accidentally encounter one at a baseball game. You leap to your feet to cheer a home run and the executive heads out the breezeway to close a deal, text a customer, or confirm an email.

Surviving—decision makers never remain under. They do not sink. They do not drown. They are giant corks. Swayed back and forth, perhaps, pulled under for a moment, maybe, but only to bobble back and start all over. Come-back kids. Decision makers never stay down for the count. They survive, they fall, they get back up. They lose hand after hand, but remain at the table. They never take themselves out of the game. If someone else takes them out of the game, they find another table. Decision makers seem never to remember their past or let it affect them in any way. They shun personal culpability.

Tenacious—they never, ever, let go. They beat a problem to death. They probe, they pry, they parry, they thrust. They keep asking questions and they keep jabbing. A presenter will get frustrated, exhausted, and lash back in self-defense, but it is pointless. The pummeling will continue until a satisfactory answer comes forth, or all the players are dead. This tenacity can compel a decision maker to dig in his heels and to stick with a decision long after facts show otherwise.

Self-Confident—they can be introverted or they can be extroverted, but decision makers are confident. Firing or hiring, acquiring or divesting, they are confident. They are confident that they make the right decisions—period. If they possess those inner voices of indecision common to others, it is not detectable.

Three baseball umpires find themselves in a bar. The first umpire raises his drink and shouts, "Some's is balls and some's is strikes, but I calls 'em as I <u>sees</u> 'em!" The crowd laughs.

The second umpire, not likely to be outdone, stiff-arms his frothy mug and retorts, "Some's is balls and some's is strikes, but I calls 'em as they <u>are</u>!" The crowd cheers.

All eyes shift to the third umpire. He hesitates not for an instant. He hammers one fist into the table and slings his beer up with the other. "Some's is balls and some's is strikes, and they ain't neither 'til <u>I</u> calls 'em!"

Decision makers do not lack self-confidence. To the decision maker, his call is not supposition or appearance, it is reality. This confidence can self-deify a decision maker. "I said it and that's the way it is."

I was at an executive retreat in Barton Springs, Texas. Martin, the president of one division, recently had implemented a new strategy for his division. The CEO was skeptical, having data that the strategy was not working. "How is the new strategy going, Martin?" he asked in a public forum. "It will go well. I implemented it," was the immediate, self-confident retort. Now, the question was in present tense, but the answer was in future tense. Decision makers do not know there is a difference. The data said the strategy was not working, but the response was, "Never mind reality;

never mind the data; never mind anything else. It is going well because I implemented it. Now, what else is there to know?" "They ain't neither 'til I calls 'em!"

To a decision maker, the present tense is all there is, albeit acknowledging that the present is uncontrollable and unpredictable, but titillatingly dynamic. In today's business world, decision makers do not know if they will even be around for the future, so there is certainly no reason to fuss over it. They do not need future tense. Do they need past tense? Do they worry about the past? Why? What good is the past? They have never heard of George Santayana's, "Those who cannot remember the past are condemned to repeat it." Decision makers totally ignore their past but, unfortunately, not yours. It is present tense with them, all the way.

I scheduled a nationally-known radio celebrity to speak at a large management meeting where I was host. In his presentation, he related that when he started the station, now a giant corporation, that there were few listeners and the forum was not catching on. (It was one of the first stations to primarily cater to women.) This continued for two or three months. What to do? The owners were recommending cutbacks and a modification of format, maybe cutting losses altogether. This celebrity took his own money and invested heavily in the station. Instead of tightening the belt or reducing programming, he did the opposite. He threw a lavish party at the end of the week, invited every important executive in the area to come by after work for a celebration, every potential sponsor, all the employees, and every salesperson. He toasted all the new (nonexistent) sponsors and contracts, and the (purported) new listeners now tuned in to the station. In reality, there were no new sponsors, no new contracts, and few new listeners. He was toasting the desire, not the reality. "We're smoking hot this week!" he announced. (Smoking, yes. Hot, no.)

A low audience level continued the next week. But, again, Friday afternoon, another big celebration of the "successes" that, in reality, had not occurred. "We're smoking hot this week," he told them again.

Another week went by.

"We're smoking hot this week."

Another week.

"We're smoking hot this week."

And the smoke caught fire. No one wanted to be left out. Sponsors jumped on the bandwagon, new contracts came in. Listeners tuned in. Before long, the party was for real.

Now, as a technologist, you may label this a gimmick, a come-on. Not to a decision maker. Like the umpire, it is whatever he calls it.

Fear of Failure—riding beside all this self-confidence is a Siamese twin, a pale horse: the fear of failure. To fail is the unpardonable sin. There is no absolution for failure. It keeps decision makers up at night.

This fear of failure, attached to self-confidence, emits a spectrum of signals. Decision makers may avoid open discussion and always work everything in secret.

They may be a public bully, but privately approachable, or vice versa. They may be controlling and monitoring, and require that they be involved in every detail. They may judge the work of others as never good enough, never enough, and never quite what they want. "Bring me another rock. I don't like this one."

Dynamic Tension—visions, yes, pipe dreams, no. Decision makers have their own sense of balance. Not a balance regarding home and work, or self and company, but an emotional balance of a dynamic sort. This balance permits neither giddiness at success nor despondency at failure. For example, in an evenly matched tug-of-war game, neither team moves. All the players sweat and strain pulling on the rope, but neither team moves, though huge amounts of energy are consumed. In a decision maker, it may look like there is no reaction, no emotion, but a lot of energy is being expended to keep that rope balanced in the middle.

There is a revival in popularity of an exercise program using the techniques of *dynamic tension.* No weights, cables, or pulleys are used. The arms and legs are pitted against each other in a dynamic strain, the one pulling, the other pushing, with no motion. Decision makers know dynamic tension.

Decision makers seem never to experience the *thrill* of victory or the *agony* of defeat, because their realism always keeps them *dynamically* neutral, like the tug-of-war. Increase the profit by 50% this month and they expect the same, and more, next month. Win a new contract today and they expect two tomorrow. No thrill of victory, because they always see the next obstacle.

No agony of defeat, either, because they always keep coming back. They are never defeated. Miss the sales forecast by 50% this month—they will offer the necessary human sacrifices and continue. There is always the next hill to conquer, and they *will* conquer it. That was last month; this is this month.

The tension part of the decision maker wants unfiltered, unadulterated information coming in, usually from varying sources. Externally, this is perceived as distrust. You tell the executive something and find her requesting the same information from someone else. She may even implement competitive groups within the same organization to hedge her bet. She may want to hear both sides of every argument, always comparing responses.

Jason Wilson was CEO of a closely held company that was working on a deal in the Middle East. He had one agent employed out of Saudi Arabia. He had another agent employed out of Lebanon, working through a US visa. Both agents were independently marketing to the same customer. Wilson never let the agents meet and never discussed the one with the other. As you might guess, it caused confusion to our customers, consternation with our agents, and dissatisfaction with our marketing department, but that is the way it remained. Wilson felt he could only get good information by having competing sources. He was the decision maker.

Autonomous—decision makers do not organize democracies. They may seek counsel; they may query opinions; they may take a vote. They may appoint committees and chairpersons. They may form teams and designate leaders. But, make no mistake, their kingdom is an autocracy. Decision makers decide alone. Ostensibly, they may pontificate that "in the multitude of counselors there is safety," but, in reality, they make the decision on their own. They may blame others, give credit to others, but they make the call, the *only* call. They are decision makers.

This is not a comprehensive list of decision maker characteristics. By no means do all decision makers even have all of these characteristics, nor, necessarily, to the extent discussed. These are guidelines to understand some of the qualities that exist in decision makers. Look for these traits at all levels of decision makers. It is not the title that makes the decision maker. It is the decision maker who makes the title.

CHARACTERISTICS UNCOMMON TO DECISION MAKERS

We have discussed characteristics common to decision makers, and we should discuss, also, characteristics uncommon to decision makers. Several studies have been done along these lines.

In one analysis, a group of subordinates to chief executive officers were asked to rank the personalities of their CEO's. They responded with the usual traits discussed above, but further differentiated traits their executives did not possess. High on the list, they noted that their executives placed little or no emphasis on ensuring that everyone felt included. The CEO had no observable recognition or desire that everyone should feel, or even be, a part of the team.

The subordinates said top executives felt no need to have a consistent approach. While the executive might ask for advice in a given instance, equally, he might not. He might want to hear both sides of the equation, he might not. He might solicit counter opinions, he might not. The "might" and "might not" could alternate capriciously. They overwhelmingly indicated that external "consistency" did not seem to be a necessary part of their executive's decision making process.

Another missing element indicated by the survey was that top executives placed little importance on self-correction. In the eyes of the subordinates, there was no indication that the top executives ever used the result of their experiences for self-examination or self-improvement. Perhaps they did, but subordinates saw no indication of it. In addition, the executives seemed never to be aware of self. If they were jerks, they were jerks. No self abatement. If they were arrogant or egotistical, the executives seemed not to notice, or care. They had no need for a mirror. They had no need to examine themselves. They had no intention of backing up.

Another trait commonly missing was that they seldom involved themselves with issues external to their fiefdom, that is, they never worried about what was happening to other organizations outside their own select sphere. Every executive was "an island, entire of itself," even when their organization was a piece of the continent and part of the mainland. Here, the survey meant that the executives focused on their own realms, their own part of the company, their own designated domains. If organizations were ships of the line, the executives were captains only of their own vessels. They would not necessarily cannon a sister ship (although accidents do happen), but neither would they form an armada, except when it was demanded as part of their jobs, or they directly benefited. In other words, they minded *their* own business.

In another test, executives were given advisors who, unknown to the executives, had been segregated into two types. One type would provide the executive only with information the executive already knew, albeit provided in different words and disguised form. The other type of advisors would provide the executive not only with information the executive already knew, but could also provide new knowledge, knowledge about the problem that the executive did not know prior. Again, all of this was mixed together and communicated in disguised form. Executives overwhelmingly chose the latter type of advisors, the ones who "told them something they did not know already." This runs contrary to the common stereotype that high-level decision makers want yes-men. It seems this is not so and other factors must be responsible for this perception of wanting yes-men.

Do women executives respond differently than men? The discussion above mostly used the masculine pronoun to avoid grammatical confusion, but do all these traits as easily apply to women executives as to men? Apparently, yes. When men and women compete for high-level decision making, they compete in the same arena. There are "good old boy networks" and there are "atta-girl societies." It is the decision maker characteristics we are defining, not the gender.

In discussing these traits of decision makers, there may arise the perception, incorrectly, that we have painted a not-so-nice picture of non-technical decision makers. Not true. Such interpretation entirely misrepresents the purpose. There is no intent here to assign moral value to behavior. And even if we did, not all decision makers have all these characteristics or to the degree discussed. We have simply referenced studies that point out characteristics common and uncommon to decision makers. Do some wizards have some of these traits, also? Of course they do. But, as a group, wizards would not be described adequately by this specific set of traits. Decision makers are. The two groups, wizards and decision makers, think and respond differently, and the gap must be understood and bridged if the wizard is to become a persuasive force.

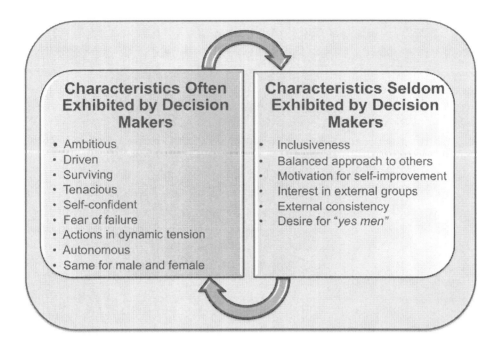

Characteristics Often Exhibited by Decision Makers	Characteristics Seldom Exhibited by Decision Makers
• Ambitious	• Inclusiveness
• Driven	• Balanced approach to others
• Surviving	• Motivation for self-improvement
• Tenacious	• Interest in external groups
• Self-confident	• External consistency
• Fear of failure	• Desire for "*yes men*"
• Actions in dynamic tension	
• Autonomous	
• Same for male and female	

Figure 2-1: Characteristics of Decision Makers

HOW WELL DO DECISION MAKERS UNDERSTAND TECHNOLOGY?

Most decision makers do not know technology to any depth. I have met only a few decision makers that do. Anthony is one of them. Anthony is, himself, a world-class genius wizard because that is his background. A few years ago, he left his position as head of a nationally-funded research facility and joined corporate America. Currently, he is an executive at a Fortune 100 company. Anthony has an odd habit. If you talk technology to him, he will ostensibly denigrate his own knowledge and abilities just before slicing you into little pieces. He will say, "I am not an expert in this field (he is) and I am only peripherally familiar with this technology (he is not), but what you just said is ... [read absurd]." Perhaps in his dual decision making/former technologist position, he feels that purported self-abasement makes him a better decision maker. I certainly never thought so. Anthony is given the title of corporate vice president within his company, but he still wears the cap of the technology wizard. No matter how different *he* thinks he has become, the other decision makers still view him as a technology wizard. *Real* decision makers know his background and weigh his advice accordingly. He is an exception, but even in his case his (business) decision maker title of vice president is known to be for show.

Neglecting a few like Anthony, decision makers usually are not technical or, at least, not technically deep. Most do not understand the technology details, although they rarely admit it, especially not in an open setting. Decision makers (like wizards) are loathe to admit shortcomings. An executive decision maker may say to a technology presenter, "I don't have a clue what you're talking about." Do not take this as a confession of any shortcoming on the part of the decision maker. She means that the presenter (do not let it be you) did not understand the material sufficiently to present it adequately. And, this is, most frequently, precisely true—the technology presenter is at fault. If you, the wizard, understand adequately the technical material, you should be able to make it understandable to an audience of non-technical decision makers. The fault is that of the wizard. Usually, it is not the wizard's lack of technology knowledge, but a serious lack of the technology persuasion skills and a serious lack of what should be presented, and how—something we will correct.

An executive may remark, "I am not familiar with that particular subject." This does not relieve you, the presenter, of any obligation. This executive will make a decision, but it will not be the decision you want if she does not understand what you are talking about. She will not make a decision in your favor unless she is convinced, and it is your job to convince. The onus rests upon the technology presenter, not the audience.

Having said all this, do not underestimate the technology knowledge of a decision maker. Some of them, like Anthony, have come up from the ranks of wizards. Some of them do have technical and scientific degrees. And some of the stuff you present is not exactly rocket science either, so the mismatch may not be as great as you think. Your job, as a purveyor of technology, is to make correct statements that are understandable to every decision maker in the audience. That means that you must know your audience, your technology, and what your audience wants to hear.

By and large, though, the decision maker will not have sufficient technology knowledge to bridge the gap, and you must do that. This technology gap is becoming more and more of a problem for the technology presenter. When I began in this business thirty years ago, it was common in industry and government to find top decision makers who, themselves, had risen from the ranks of formally educated, practicing technologists. While this is still true in some startups, the balance has overwhelmingly shifted. What caused this?

Over the last few decades, technology became broader, deeper, and more complex—specialization isolated technologists even from each other. As companies grew, they diversified, moved into *services* that had non-manufacturing and non-technical priorities. The government demanded diversity so with the shortage of technology degrees, individuals were quickly put into (technology) decision making positions, individuals who had little or no technology education. In addition, companies swiftly moved technology outside their own organizations and outsourced.

Today, the top- and middle-management of technology companies, as well as government organizations that fund technology, are saturated with MBA's, degrees in political science, administration, finance, communications, marketing, and other non-technical disciplines. These new decision makers frequently possess little or no formal training in technology.

Even decision makers with formal degrees in technology may be decades removed from actual practice. Technology has changed enormously during that time. The technologies of two decades ago are not the same as those today. Think of all the new technologies in just the last decade in cyber, nano, digital, fiber, and virtual, as examples.

The net result is a broadening dichotomy and a deepening chasm separating the wizard from the decision maker, at least as measured by technical acumen. It is imperative, then, that a technologist understand what drives the decision maker, what factors influence her most and what information she deems cogent.

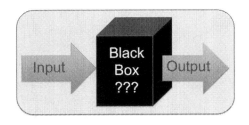

Beginning physics classes often solve the problem of the "black box." No information is given as to what is inside the black box, but the student is permitted to make tests—shake the box, use a magnet, weigh the box, measure the dimensions, hook up probes to the box, all to determine what might be in the "black box." Think of the decision maker as a black box. As technology wizards, we do not know what is inside the black box (heaven forbid), but we can observe it, make tests on it, run experiments, and see what happens. If we figure out how it works, we can know how to get the output we want. It is not necessary to know *why* the black box behaves in such and such a manner, we just want to be able to predict how it will do so for a given input.

I raise two boys. They say, "Dad, you know Mom pretty well, don't you?"

I say, "No."

They say, "But you always seem to know what Mom will do in a given situation, don't you?"

I say, "Yes."

They say, "Well, isn't that knowing Mom pretty well?"

I say, "No." I continue, "Over the years I have observed your mother. For any given situation, I know almost invariably what she will do. That does not mean that I have even the faintest idea as to *why* she would do that."

You do not require to know *why* a decision maker would act in a certain way. You need know only what that action will be. Like the black box, you need only predict output versus input. How do you do that?

We began already. We observed the general characteristics of decision makers. Let us now probe this black box and find out how it receives, processes, and reacts to input information. We need to understand how what goes in relates to what comes out, input versus output and what happens in between. *How* does a decision maker make a decision? We do not need to understand why.

We investigate by looking at some real examples of decision makers. We examine the input and the output and, by already knowing the outcome, we understand the thought processes that went into the problem and can evaluate.

HOW DECISIONS ARE MADE

Tiberius Caesar ruled a Roman empire that stretched from Persia to Spain. He set up hierarchical governments throughout with middle managers everywhere. Judea, a division of the empire, held two important seaports, Jaffa and Caesarea, that pivoted key Roman trade routes. One middle manager of this internal territory was Pontius Pilate, titled governor. Pilate's direct superior was Herod, a vindictive potentate who spent most of his time on the coast or in his mountain retreat. (It is a hot climate.) Herod reported to Tiberius. Pilate was a Roman, but his job was governing a diverse culture. His management style was to work almost entirely with the ranks of his peers and immediate subordinates. He had limited knowledge of or interaction with his constituents. Pilate, as a decision maker, would be characterized as staying at a high level with little regard for details.

His constituents, recall, were not Romans. They bring before him a man for whom they seek the death penalty. The charge is cursing Yahweh, the Jewish god. This is a serious infraction of religious Jewish law, but of little interest to a Roman who has many other gods to worry about. From a strictly legal viewpoint, the Romans cared not a whit about cursing non-Roman gods. Being subservient to Rome, the Judean leaders could make charges but could not implement the death penalty. Such a sentence required the decision of the governor, Pilate. The Judean leaders bring Jesus to Pilate requesting the death penalty for this non-Roman crime. Pilate is the decision maker.

Now, Pilate's wife tells him about a disturbing dream she had the night prior. Because of this dream she warns Pilate to have nothing to do with this righteous man, Jesus, who was known for his ministry to the proletariats and bourgeois and his disdain for the Jewish leaders, probably coinciding with Pilate's own opinion of them. On the first encounter, Pilate avoids a decision and sends Jesus to Herod, the territorial regent and Pilate's superior. Herod was visiting in the region, had heard of Jesus, and wanted to see Jesus perform some miracles. Pilate thought that this act of judicial deference would place him in good standing with Herod, and it did. Herod interrogated Jesus and returned him to Pilate for trial. "Your call," he said. Prior to this, the relationship between Pilate and Herod had been difficult, but

Herod correctly saw this as a good-will gesture on the part of Pilate and the start of a political symbiosis.

So, when Pilate enters the courtroom for the actual decision, he has four things affecting him. One, the personal bias based on his wife's dream that it would be best to release Jesus. Two, he had made political gain with his boss, Herod, and wanted to keep that—he could not risk having the Jewish leaders, his constituents, launching a complaint that might reach Herod or even Caesar. Three, his own sense of justice, whatever that might be, based on his background and education. Four, the facts in the case.

Pilate heard the prosecution's arguments against Jesus and ruled the Judean case to be one of esoteric Jewish law and not worthy of the Roman death penalty, stating directly, "I find no fault in the man." He showed integrity. He was going with factors one, three, and four.

The Jews tried another approach. They told Pilate that Jesus had claimed to be a king and that would amount to treason against Caesar—insinuating that Pilate would be in jeopardy with Herod and Caesar if it were advertised that Pilate released a Roman traitor. This argument appealed to Pilate's improved relationship with Herod and launched a certain fear of reprisal should such an accusation find cause in the ears of Herod, or worse, Caesar. In a flowery show of political rhetoric, Pilate washes his hands and releases Jesus for crucifixion. He was going with factor number two.

Every presentation before decision makers has at least these four character-istics—personal influence, political ramifications, assessment of value, and facts and data. Every decision maker comes into the meeting with prejudice, influ-ence, values, and assessment. Before the technologist ever says anything, there are weights already positioned on the balance scale of decision. As a presenter, one must recognize these influencing factors, understand them prior to the presenta-tion, and work to persuade the decision maker—often to turn her away from a pre-established position or bias. This is difficult because you may not know or correctly analyze these biases. You must learn how to do your homework, how to read reac-tions *in situ* and how to work solutions in real time. You cannot be successful just blindly coming into the meeting and presenting technology like it is a science class. It is an art.

In examining Pilate as a decision maker, we understand that the presenters changed their recommendation to make it relevant to Pilate. The first charge, if not one of technology, was certainly one of technicality. The revised charge was one of treason. Pilate was a political appointee and that was the angle they successfully attacked. He had biases and a personal predilection to release Jesus, but they found an argument that persuaded him. He gave them the decision they wanted, if not for the reasons they wanted.

This is a key analysis that the technical presenter must take to heart. The wizard must understand the influencing factors that drive each decision maker. What are the biases, the level of understanding, and the background? A medical technician may prepare a presentation with reams of good experimental data and dozens of charts, and these would satisfy a researcher. They may be inadequate, or even inappropriate, to present to the administrator.

The decision maker has manifold considerations.

It was a blistering hot day in Riyadh, Saudi Arabia; the sun was scorching by mid-morning. I brushed sand particles off the back seat of the auto before I sat down. I started my computer and began to review the presentation I had prepared. My agent was sitting in the front seat with the driver, exchanging pleasantries. I worked until we stopped. I looked out the back window. It appeared the driver had let us out at the back of a large, nondescript, concrete facility that was possibly under construction, boarded walkways blocking any indication of the overall plan. Guards appeared from a portable building, checked our identities and led us through a maze of makeshift plywood walkways. We passed through the metal detectors, unzipped briefcases for inspection and walked, rather swiftly I thought, down long halls and through adjoining rooms. We stopped in a moderately sized anteroom crowded with neatly stacked documents and were told to take a seat. In about five minutes, the door opened and a lone individual entered. It was a member of the Saudi royal family, a prince and commander of the Royal Saudi Air Force.

We rose, introduced ourselves and exchanged the standard protocols. But he quickly got to the point. "What do you have," he said.

It was up to me to make the presentation.

I was there to explain a unique satellite technology we possessed. I wanted him to invest 50 million dollars, commit to a three-year contract and take delivery a year from now—*if* we were successful in launching the satellite.

Now, the uninformed think that Saudi Arabia has more dollars than sand, that they want to buy everything, that all you have to do is ask the right person and then show them where to unload the money truck. The uninformed have no knowledge of Saudi Arabia, or the Middle East. Otherwise, they would know:

a. People who have more dollars than sand want to keep their dollars and sell their sand;

b. People who have more dollars than sand are infested with people who want their dollars and not their sand;

c. People who have more dollars than sand also have more hours than sand, where time is measured with a calendar, not a watch.

This was a massively difficult situation—a foreign country, people who had different values, a high ticket item, complex technology, long lead items, technical risk, government regulations to satisfy, very little time to present and little knowledge of what competing factors might be in the mind of the prince. For a start, he had a similar bid from the French that was conspicuously left sitting on the table behind me.

So, you see the problem. It was easy to analyze the decision making of Pilate. Anyone could do that. Up close, in reality, it is massively difficult.

A startup company approached me with a new concept for electronic ink, wanting my company to partner with them. I was the decision maker. I listened to their presentation for an hour. It started bogging down with technology details. I could tell they thought I did not understand the technology because they kept wanting to tell me so much about it. Not so. I did understand. I had read their material the night before, done my homework, and understood exactly what they were proposing. What was I doing, then?

I was thinking. It sounded like a great invention, but was there sufficient intellectual property to protect our investment? Would the technology stand up against the "prior art" clause if their patents were challenged in court? Was most of the profit in the US, or was it in countries where the patent protection was less secure? Were their projected research and development costs realistic? Was there a window of opportunity or were we already too late? Did this invention leap-frog the competition or just keep us in the game? Did I have sufficient available resources to pursue a successful partnership? What was the competition doing? Did we have the technical wherewithal to bring the idea to market? Were our pockets deep enough? Did we have a champion, a leader for development? Was it a market we knew? How far was this from our comfort zone? How close to our sweet-spot? How and where would we manufacture it? Those were my thoughts as a decision maker. This incessantly technical wizard was not doing much to help me with answers.

The decision maker is consumed. Did I mention that I was hungry and wanted to get some lunch, also? That I had to finish a report for my boss and I was already late in getting it to her? When you are on the other end, when you make a presentation, a technology presentation, there are a multitude of competing challenges facing the decision maker and you must learn to detect, deflect, and orchestrate them.

The decision maker is working a decision matrix and the technologist must get into that matrix to be successful. A technologist must bring insight to the things that matter to the decision maker. But first, the technologist must know what these things are. Many of the elements of the decision matrix are foreign to the wizard engineer or scientist. They are not in the wizard dictionary. We are beginning already to put them into your internal glossary.

Most technologists are focused on their own results and their own research. They are rightfully proud of all their fine work and think that decision makers want to hear every detail. Being such fine work, how could a decision maker not want to hear it all? Easy. As a presenter, shift from thinking about yourself, your needs, and your results, and move to the matrix of the decision maker, his needs, and his results.

Factors Affecting All Decision Makers

- o Personal Factors
  - • Attitude, family, disposition, emotions
  - • Competing thoughts, more pressing issues
  - • Value, financial gain, profit
  - • Politics
- o Personal gain, relationships, advancement
  - • Tit-for-tat
  - • Fear of failure
- o Baggage
  - • Prior experiences, successes or failures that bear on the decision at hand
  - • Commitments, promises, regrets, revenge
- o Facts
  - • Business case, capital, profit, margins, business model
  - • Available personnel, technology champion
  - • Ready customers
  - • Sales and marketing
  - • Competition, window of opportunity
  - • Barriers to entry, legal protection, competitive fences

BEHIND CLOSED DOORS – WHAT REALLY GOES ON IN THOSE MEETINGS?

What is it like, really, to make a technology presentation to senior-level decision makers? What transpires in those meetings? What are they thinking? What happens before I enter the room and what happens afterwards? Do they discuss my presentation? Do they discuss me? What are they thinking and doing while I am presenting? What things am I doing right? What things am I doing wrong?

These are questions that all wizards have. Brace yourself.

You were told last week that you must make a presentation before senior management. Since then, you have spent hours working on your presentation. You are as prepared as you know how to be. Your presentation is at three o'clock. It is now one o'clock. You wander past the entrance to the conference room. You understand that the decision makers have been in session since early morning. The door is open. They are returning from lunch. The chief operations officer (COO) goes in and sits in his customary place followed quickly by the various vice presidents of the organization. The head of engineering rushes in behind the chief financial officer (CFO). There are attendees who are participating by videoconference and you see their faces on one of the screens. They shut the door and go on with their meeting.

How do you find out what is going on inside? Let us imagine that, in addition to being a technology wizard, you are also a real wizard, like one of those in the movies. You have the ability to turn yourself into a little fly, to make yourself unseen, hide in the shadows, and be a fly on the wall. You glance around, see no one else in the hall, slip into a corner and transform yourself into the fly. You buzz unseen under the door and into the conference room.

There is a presenter at the podium. The presenter begins. He opens with an apology. He does not have all the data he planned to have by this time. The project has been stymied for three weeks. He offers some reasons why. Ouch! He infers blame on another organization whose representatives are apparently in the room. They are visibly displeased that the presenter has waited until this moment to foist the blame on them.

The president of the company is noticeably angry. The project has missed milestones already, is falling further behind, and costing more money. The speaker offers a vague projected date for completion. The president fires invectives at responsible individuals. Two underlings pile on for a general massacre. It is getting tense. The speaker is requested to continue.

And, continue he does. Perhaps the speaker knows what he is talking about, but, if so, only he. The rest of the room is glassy-eyed. Details, details, details. The presenter goes off on tangents and rabbit-trails and has to be brought back several times. He is well past the scheduled thirty minutes. He is attempting to make up, in the presentation itself, for lack of progress on the project. But this is the presentation and not the project, so that is useless. Three executives are typing emails into their Blackberries, holding them below the table, but obviously giving the presenter only snippets of attention. The speaker is interrupted with questions for which he has weak answers. The president finally terminates the already tardy presentation and the presenter is taken off the hook and thrown back into the sea of technical oblivion. He traipses out of the room. The executives take a few minutes to discuss next steps and decide to cancel further research on that project.

The next presenter starts late, of course, but she promises to make up for the lost time and to finish her presentation on schedule. Unfortunately, she is able to do this only because her test results were negative and she has nothing to say, anyway. The executives had no prior warning of this. Her data show that the new drug failed Phase Two test trials. They ask her about an alternate plan. She replies she has not yet worked out an alternate plan of approach. It is almost as if she wants the decision makers to tell her how to do her research. No one is happy. Everyone is bristling. She is asked questions. She responds too quickly with answers that are incorrect or, at best, inadequate. At last, she is finished. The meeting remains behind schedule. An administrative assistant is sent to get you for your presentation. Oops! Time for the Fly to revert back to the technology wizard.

It is your turn, but the meeting is still running almost 30 minutes late because of the other two presentations. The secretary tells you the president has a meeting with the CEO at the top of the hour and has to leave a few minutes early to be there. You prepared material for what you thought would be an hour-long presentation. That allocation has now been cut to forty minutes. There is no way you can make your presentation in forty minutes. In actuality, you had more material than you could have presented even in a full hour. What to do? Well, give it a try.

You start talking as fast as you can to get it all in, but you are not into your fifth sentence before an executive stops and asks you to wait a minute. The president is whispering to the CFO. You do not know it, but they are still discussing the presentation before yours and one of them has a new idea. Obviously, the failure in the Phase Two test trials is still on their minds and they have not yet shifted their attention to you. Shortly, you get the motion to go ahead and continue. At last, you have their attention. Maybe. What to do?

You continue at the most rapid of pace, but the material is too complex to be grasped that superficially. Your manager is in the back of the room. The little blood vessels on his face are beginning to dilate. He starts giving you the hand signal to speed things up. You're already going at the speed of light. The decision makers seem to be elsewhere, anyway. At ten minutes to the hour, they stop you in midstream. The president leaves early to go to a meeting with the CEO. He leaves someone in charge, but your time is up. You are not even close to finishing. Failure. No decision has been made on your project.

What does that mean, no decision? It means that a decision was made – to do nothing. "No decision" is a decision, and not in your favor.

This scenario is not unusual. It is typical. It included common mistakes of most wizards and concomitant reactions of most decision makers. Briefly, then, we perform a short critique to spotlight the elements we must address in detail in the rest of the book.

There were several things you could have done to avoid failure. As a start, you could, and always should, have had a plan in place for a potential, significant loss of time. Second, you should have had a plan in place to transition from the presentation before yours, especially if you have any indication that the meeting may have been hostile. Third, it appears your material was not organized or at least not presented adequately for them to grasp it. Fourth, you did not get a decision in the meeting. You must never end a meeting without a decision.

Those were a few of your mistakes. We find solutions shortly that will make you successful.

SORTING OUT THE PLAYERS

In my experience, I find that many wizards do not know even the rudimentary elements of how a company functions. Thus, when they make a presentation, they step on toes and needlessly explode land mines. Let us take a moment to examine, at least at the highest level, how a company is organized.

At the top of the ladder are the owners. In a public company, this will be the stockholders. In a private company this can be a single individual, a group of individuals, other companies, or a combination.

The board of directors represents the company to the owners. The board has bylaws that dictate their span of control, behavior, responsibility, and representation. These bylaws were ratified when the company was put in order. The chairman of the board resides over the board. Sometimes, the boards meet in special session. Boards usually have quarterly meetings to discuss the performance of the company. The rest of the members of the board of directors are called directors, with various committees and responsibilities within the board.

The chief executive officer (CEO) has responsibility for the execution of the company's entire business. This is a corporate level position reporting to the board.

All these are functions. A single individual, for example, may be both the chairman and CEO, depending upon how the company wants to run itself.

Large companies are divided into divisions, or some similar name for the same function. At the corporate level and within each division, there are a host of functions.

The president of the corporation, or a division, has responsibility for the entire corporation or division. Along with the president are a number of people who make up his staff and inner circle. The chief financial officer (CFO) has responsibility for all matters financial—reporting, auditing, and budgeting, for example. The chief operations officer has responsibility for the day-to-day operations of the corporation or division. There may be other "chief" titles like chief technology officer.

Below the President, strictly managerial positions usually carry titles like vice president, director, manager, and supervisor.

The title is not always a rank of decision making. Banks, for instance, copiously distribute the title of vice president.

I went to purchase tires for my automobile. It was almost closing time. There was a mismatch between the width of the tires they had in stock and the wheels on my car. The young man helping me seemed not to have a clue what to do or how to do it.

"May I speak with the manager?" I ask.

"He's not in, but I'm the assistant manager," he replies.

"Oh, okay," I say.

He goes into the storeroom to check his tire inventory. I wander around and began to look at the pictures on the wall showing all the employees. I find my guy.

There is his picture and under it the title, "Assistant Manager." I look at the other pictures. What? They all have the title "Assistant Manager." Then, I see it. The fine print says, "Assistant Manager #1," "Assistant Manager #2," and so on. My guy is "Assistant Manager #5." Everyone in the store is an "Assistant Manager." Be careful about correlating titles with decision making authority.

If the company is large enough to have divisions, it will have functional groups within each division such as legal, marketing, sales, engineering, manufacturing, quality, security, maintenance, operations, and human resources, for example. There can be titles of executives within each department of each division. The subsets can continue and intermingle depending upon how the company is organized.

Now, mix these titles and functions and organizations together and you get the *vice president* of *marketing* for the xyz *division* of the ABC *Corporation*.

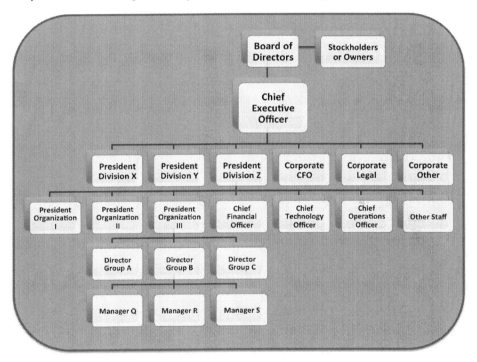

Figure 2-2: Typical Management Hierarchy

In most meetings, the hierarchy is evidenced by distance from the head of the table.

It works like this. If the president thinks your request is strictly an engineering decision, she may defer to the engineering vice-president to make the decision. It is assumed, in such case, that the decision stays within pre-established guidelines, forecasts, and budgets for engineering. Thus, for any meeting, you must establish ahead of time exactly who will have responsibility for making the decision. This is a part of your homework that we discuss later.

A dynamic problem often encountered is action from a "Wild Card." The Wild Card can be an executive who wants to torpedo your idea because, say, it competes with his division. It can be a technology consultant brought in by management to listen to your presentation who thinks your schedule is unrealistic. It can be another executive who offers up, "Hey, I know a group over in Milwaukee that can do that cheaper." The Wild Card must be forecast whenever possible, and nullified with a suitable plan of action (of which, more later).

A STEP IN THE RIGHT DIRECTION

There, you have it. Decision makers and decision making. Now that we know what the decision maker is like, and how he makes a decision, how do we ensure success? Before we start making lists of things to do and things not to do, let us establish a strategy. A battle is not won without strategy. Once we have a winning strategy in place, we can fashion our technology presentation around that strategy.

Know the Game and the Players

- o Presentations prior to yours:
  - May affect attitude toward your presentation even if unrelated.
  - Almost always undermine the schedule.
  - May preoccupy the decision maker's mind.
- o Know the hierarchy and decision making authority within the company.
- o Know who is the decision maker for your recommendations.
- o Watch out for the *Wild Card*:
  - Interference from within the organization.
  - Consultants or other technologists who argue with you on key points.
  - Decision makers not in the direct line who may exert control.

Chapter 3: Stereotype Mitigation
Harnessing Doubts and Anxieties

Life is not easy for any of us. But what of that? We must have perseverance and above all confidence in ourselves. We must believe that we are gifted for something and that this thing must be attained.
— Marie Curie

Wizards are a stereotype. No doubt about it. Inform a stranger you are a physicist, chemist, computer programmer, telecommunications technician, or engineer, and you will *feel* the reaction. Those wizard job titles advertise "introverted geek who came from another planet and plans shortly to return." My experiences have unearthed little in the way of contradictory evidence.

Consider stereotypes. I grew up in the oil patch that is west Texas. A "pumper" is an oil field worker who monitors the pump jacks, measures oil in the tanks, opens valves so that crude oil can pass through refining equipment, and then opens another valve to let the refined oil channel into pipelines for delivery. As the technology changes, new equipment is introduced and new procedures are specified by the oversight engineers, who are mostly not in the field, but directing from an office in Wichita Falls or Dallas. Engineers visit the fields periodically. Now, suppose an engineer specifies changes to the procedures or equipment. If that engineer introduces the slightest innovation that is considered by the pumper to be novel, difficult, or harder to implement than what existed already, never mind if production increases ten-fold, the pumper will issue profane denigrations that end with something to the effect of "it's another engineer's *bright idea.*" The field workers disparage the engineers who keep them in jobs. They stereotype engineers as rich, educated oil men who sit in plush Dallas offices and know little about the real workings of the oil field. They stereotype them. Is the stereotype correct? That question is immaterial.

In Texas, we joke about the "redneck."

You know you're a redneck if you get stopped by a State Trooper. He asks you if you have an I.D. You say, "'Bout what?"

Or, maybe:

Janice said she liked the Dollar Store because, "You could go there all casual-like and not have to get dressed up like at Walmart or somethin'."

In the business world, decision makers think technologists know nothing about business, and technologists think decision makers know even less about technology.

Neither view is correct, but the stereotypes persist because both opinions get sound reinforcement.

Decision makers stereotype the wizards who present to them. Some of the stereotyping is valid. Some is not. The best we can do is circumvent, rather than eliminate, those stereotypes. Therefore, in making a technical presentation, it is paramount to not only understand the decision makers, as we discussed in the previous chapter, but also to remove your wizard stereotype from their decision-making process. Whether you actually fit the stereotype is unimportant. You must eradicate the perception in the minds of the only persons who matter in the conference room: the decision makers.

Of initial importance in mitigating the stereotype is to develop confidence as a presenter. A nervous, ill at ease, uncommunicative presenter reinforces the wizard stereotype. This is a dilemma because most wizards' personalities are, indeed, terrified at making a public presentation. The goal is both to possess and to assert confidence in the conference room, not confidence in the technical content, which is simple, but confidence in the ability to present the material persuasively to decision makers.

Over my career, I have listened to thousands of technical presentations. I have tutored hundreds of engineers, scientists, and technologists. I have never met one who had no initial fear of public speaking. Most possess it forever. It is obvious in their mannerisms, in the material they present, in the way they organize the material, and in the style they use to communicate it.

What you may not know is that everyone, not just scientists and engineers, have trepidation when speaking or performing publicly. Entertainers, actors, and even the most professional of technologists has doubts and anxieties about the ability to perform to the level required.

A nationally-known commentator once confided, "Every time I step on stage, I am terrified." This is hard to believe because he exudes master confidence and control in his presentations. Yet he lays claim to the same reactions as we normal people. Those of us who are called on only occasionally to make presentations, he has the same reactions as we do—sweaty palms, dry lips, shaking hands, weak knees, queasy stomach, and a pounding heart—all symptoms of that phobia known as stage fright, possibly the most common plague on Earth.

I attended a training seminar at the state capitol. A deputy chief of staff was scheduled to present to our audience of 300 trainees. She sat on the stage prior to her presentation: attractive, mature, polished, and poised. She looked confident. Certainly, she had nothing to fear from us. When she got to the platform, this is what she said. "I worked for a number of years as a reporter in Austin. I interviewed every prisoner on death row—sitting directly across from the most reprehensible criminals in the state. I sat witness to every execution. In none of those interviews or executions was I as nervous as I am now, facing you." That was her statement. It was a strong one. She feared speaking to us more than she feared interviewing a heinous murderer.

It is called stage fright and we all possess it. It is not rational, but it exists nonetheless. Every speaker, especially every infrequent speaker, experiences stage fright.

There is no single, good word for the plague. "Stage fright" is not the optimum word since it clearly does not require a stage. The fear can come from just sitting in a room and having to say something about anything. It can even be an informal setting. Some refer to it as a fear of public speaking. But it does not require speaking, either. Even actors who have no speaking part get stage fright. It does not even have to be very public. Only one or two listeners can be sufficient to bring on all the symptoms. There is no single good word for it and there is little rationality for the condition, but we all experience it.

I am on the Board of Regents for Midwestern State University. In this position, one of the proud honors is to be on hand to pass out diplomas at graduation. In the ceremony, we regents are on a wide platform at the front of the auditorium. Each seated row of potential graduates is sequentially queued to form a line leading to the provost. The provost is off stage left. The provost calls each graduate's name, the graduate steps onto the platform, and walks across to stage right to receive the diploma. As they mount the stage and walk across it, most graduates are smiling, a few are serious, some are looking at parents, and some are raising their hands with the victory sign. I am standing about a third of the way across. I stick out my right hand, which they shake, I congratulate them, and then I give them their diploma with my left hand, which they then take with their left hand and move on to the president of the university who has his picture taken with them. They move off, stage right, and go back to their seats. Now, how difficult can that be?

All each graduate has to do is stand in line with fifty peers, walk across the stage, shake hands, and receive a diploma. They practiced this only yesterday. Yet, from the biggest football player to the tiniest cheerleader, from the oldest grandmother to the youngest prodigy, African, Asian, Latin, or American, every one of them—every last one—has a palm soaking with sweat. They all have stage fright. (And we have a box of clean wipes backstage.)

Absolutely everyone has some amount of stage fright. It may not look like it, and they may not act like it, but they do. Those who look calm and collected have learned to control it. Also, they have reduced its intensity. The biggest bundle of fright comes from antici-  pation. Once you are in position to "do your stuff," most, if not all, of the stage fright goes away. In any case, you can train yourself to make it happen that way.

Stage fright is necessary, even required, as it releases the adrenaline to speed up your thinking, generate your passions, and make split-second mental connections to expound your wherewithal as a technologist. Those persons who have no trepidation whatsoever, if they do exist, do so because they are too full of themselves to know fear; they will not be successful at technology persuasion because it is all about them.

Perhaps there are those, also, who are so experienced that they have no stage fright at all. I have never met them, but perhaps they do exist. Then, that experience prevents them from improving. It actually becomes a barrier to persuasion. There are no sports stars so great or no performers so fantastic, but that they sometimes lose, get in a slump, or reach a plateau. They can improve. I remember a few years ago hearing that the late, great Luciano Pavarotti had apologized to an audience because he, and they, felt that his performance of that evening was not at his best. He was tops in the business, but not on that day. Josh Hamilton once won the American League batting championship with a .359 batting average. Over the next two weeks, he played in every game of the World Series. He had two hits in twenty times at bat, an average of .100 for the entire World Series. It happens to everyone.

You must keep pushing and striving to improve. You cannot improve unless you feel the need to improve. You need that fear, that stage fright, to examine, to dig deeper, to ensure adequate preparation, and to hone your skills.

What you do not need is unreasonable, misdirected, or uncontrollable fear. People who generate an excessive amount of fear are often overly concerned about the slightest flaw. Put the right technical presentation together and two minutes into the presentation you will forget about fear. However, that adrenaline will still pump and you will give a solid, productive presentation.

Many years ago, I met George Meyer. He was a very old man at the time. George laid substantial claim to being the driver of the getaway car at the infamous St. Valentine's Day massacre. He wrote a book about his role as devil-driver for the Capone gang. On February 14, 1929, Al Capone's Chicago syndicate murdered seven hoodlums of Bugs Moran's rival mob. Afterward, the assassins jumped into the getaway car, apparently driven by George, and sped away. No one was ever tried or convicted for any of the murders. When I met George, he was involved in a church ministry, trying to redeem value out of his thirty-one years in Alcatraz, Attica, and Dannemora, and a life spent in crime. He told me the problem was that although he was a popular church speaker, no one wanted to hear about all the good Christian things he was now doing. They wanted to hear the guns and guts stories of Al Capone, Frank Nitti, and "Greasy Thumb" Guzik. (That was certainly what I wanted to hear. I do not remember a thing he said about Christianity, but I remember every word of the guns and guts.) And, that is where we are in this discussion. We are tempted to go immediately into the guns and guts of how to make a presentation, but find we must first learn how to stay on the stage so we can actually deliver our guns and guts presentation. We have to work on fear a little more.

There are several causes of stage fright that are peculiar to the technologist. We need to resolve those because they arise from the way in which technologists are internally wired. We will examine those factors, understand them, and then work to mitigate and overcome them. Most of the counterproductive elements of stage fright will then disappear. The productive, desirable elements will remain. After we eliminate the undesirable elements of stage fright, we can go on to the other elements of a presentation. Here is a list of things that produce stage fright, especially in a technology wizard, and what to do about them.

Now, we are at the beginning of a rather long list of things that cause stage fright, particularly in technology wizards. It is quite a lengthy list, but that is because the chasm is broad and there are a lot of pillars that must be put under our bridge if we are to cross over to the side of the decision makers.

APOREO

Aporeo (ἀπορεω) is an Attic Greek verb for which there is no direct English equivalent. It means *not knowing what to do*—a common enough occurrence so it is surprising that no such English verb exists. Nouns like "quandary," "perplexity," and "helplessness" give the idea, but there is no equivalent verb. When given the task of presenting to decision makers, most technologists would find that *aporeo* clearly expresses their dilemma: they do not know what to do. And there is no one to tell them. Hence, the purpose of this book.

A number-one cause of fear in technology wizards, other than the presentation itself, which we cover later, is that they will be asked non-technical questions for which they are ill-prepared and not knowledgeable. So, they start their preparation with *aporeo*, never quite sure what to do, and end up woefully unprepared. It is a punishable sin. You simply cannot do well if you have not prepared well, meaning prepared correctly. The usual problem is that the technologist is well-prepared to present technology to peers, but not to present arguments to decision makers. Wizards know there is no absolution for inadequate preparation, but they do not know what to do about it for an audience of decision makers. You must be adequately prepared for persuasion. Otherwise, *aporeo*.

For the technologist, correct preparation is mandatory. You can whine a Yankee brogue, you can drag a southern drawl, you can even have an innate speech impediment. You can be a walking lexicographer or stumble over six-letter words. Your thoughts can be as complex as Einstein's or as ingenuous as a fifth grader's. You can speak in Faulkner paragraphs or Hemingway clauses. With all of these, you can still exhibit powerful, successful technical persuasion. But you cannot do it if you are inadequately and incorrectly prepared for the decision makers.

First, you must know what you do not know. In other words, you must know before you present where the data are incomplete, in error, or missing, and not have to wait for a decision maker to point it out in the meeting. You must know what is lacking, what is known, and what is not known. You must know the validity and scope of your projections and statements. There is nothing more fatal than to answer a question and then have the decision maker negate your answer with facts you should have known. Or, have a decision maker ask reasonable questions for which you have no adequate response, and did not even consider before she brought it up. You are the expert, but you must play the game in their arena, not yours. You do not have home field advantage.

Prepare *correctly*. You do not have to be the smartest person in the room. You do not have to be the company genius. But, on that day, on the subject you are presenting, you must know what you know, what you do not know, what you should know, and what you cannot know. You must be prepared for the audience. Being correctly prepared will go a long way toward taking away your stage fright. Like the Boy Scouts say, "Be prepared." If you are prepared correctly you will possess an inner confidence to manage the stage fright.

THE FASTER GUN

Another critical factor that causes intense stage fright in technologists, is a technical version of the first factor. The first factor has more to do with being unsure of the non-technical content. The second one is being unsure of the technology. It is the fear that you will say something incorrect or "stupid" that will be caught or contradicted by a more knowledgeable technologist. It would be wonderfully ideal if you were the most knowledgeable person on every element of the presentation, but that will not happen. Projects are too broad and science too specialized for it to occur. Almost every large project now is done with teams. It is nonproductive, wishful thinking to expect to be the sole holder of the knowledge keys. As we all learned from old Wild West movies, there is always a "faster gun." And there will always be members of some audiences who do know more about specific parts of the subject than you. What to do?

If you know ahead of time that an especially knowledgeable person will be at the meeting, spend some time with her prior to the meeting. Dry-run the presentation with her, if possible. We discuss this technique in depth in Chapters Twelve and Thirteen. Let the expert point out your errors and the weaknesses of your presentation and then work on those areas. Sometimes, it makes sense to ensure this expert attends the meeting to help give you cues or interject things into the presentation. It may give your presentation validity. Sometimes this is good and sometimes it is bad. It depends on the personality of the expert. If the expert can play the role of a silent ally, then better to have her there. If she is a cross between Chatty Cathy and Annie Oakley, then do not ask her to attend, but tell her you will reference the help and guidance you received from her. When given the choice, I choose the latter as it permits better control of the presentation and avoids the potential injection of technology comments not apropos to your train of thought.

If the expert has been asked to the meeting by a decision maker and you cannot, offline, resolve your differences with that expert, then do your best to avoid a frontal attack. If you fear that the expert will attack you in the meeting, then plan accordingly.

In *Zen and the Art of Motorcycle Maintenance,* Robert Pirsig suggests that when facing a charging bull, the "other expert," you have more options than just standing there and being gored. You can "throw sand" in the bull's eyes. You can admit that you are not competent in certain areas and therefore choose not to address them. I strongly object to this technique. No one knows everything. There is nothing useful in pointing out every area where you are deficit. For most individuals, that would take more time that is allowed in any meeting. Just stick to what you know and stick to the facts and the technology.

A better approach is to defuse the bomb. Do this promptly, before it has time to explode, before you get interrupted by the expert and get in a shouting match. Within a minute or so of starting your presentation, you interject that you have met with the expert and you two have gone over all the details and you agree in some areas, but still disagree in other areas. Reinforce that you will be giving your position and that you will point out where you disagree with her. Nevertheless, reinforce that you are here to present your position and that there is not sufficient time for both you and the expert to discuss with the entire group what you have already discussed offline. You tell them you will be happy to take actions afterward or come back and discuss more details, but that this presentation is to present your point of view. This defuses the bomb early on before it explodes. It also relieves you of this part of the stage fright. You know ahead of time what to expect and you have sidestepped it. Your internal reinforcement should be, "Of course, there is another expert in the room. Of course, I have discussed this with her. Of

The Faster Gun comes in many forms; More Technical Knowledge, Recognized Expert, Consultant, Ego Gone Wild, Competing Project, Personal Vendetta

course, there is disagreement. Now here is what I have to say." Sidestep this fear and do not take it head-on.

What this bomb-defusing approach also does is gain the confidence of the audience. Now, they are prepared to hear what *you* have to say. Now, they will defer to your presentation and give you your day in court. They will listen to the points of technical opposition and be ready to make a decision. Now *you* are in charge. But do not expect to finish without a struggle. Your best tactic, when interrupted by the expert, is to repeat what you said earlier, that there are disagreements, and then proceed with your presentation as if nothing were said. Do this a few times and the decision makers usually will intervene on your behalf.

Do everything you can to avoid direct confrontation in front of the decision makers. There will not be sufficient time to do battle and neither of you will win. Worse, you will both lose. Do not make the mistake of ignoring the expert prior to the meeting and coming to the meeting hoping you will not have to do battle. Egos can run high and they are best soothed offline.

None of this addresses whether the expert is truly an expert or not, or whether the expert is right or wrong. The expert can be dumb as a post, but your assessment is unimportant. Remember, first of all, that most wizard technologists feel superior even to peers. Secondly, the only opinion that matters in the conference room is the opinion of the decision maker. If the decision maker thinks Cathy is an expert, then Cathy is an expert, and you need to recognize and accept Cathy as an expert.

But, remember, expert or not, if you draw your sword to parry a technical thrust, the duel is on. The audience will go glassy-eyed. They will be confused and leave without reaching a conclusion. Do not duel in the presence of decision-makers. Of all this, more later.

WHAT IF I MAKE A MISTAKE?

A fear that persists throughout your career as a technologist is that you will do or say something that is incorrect or perhaps outright wrong. What if that fear is realized and you really do make a mistake? Strangely, the more important you are as a technologist, and the higher you rise in your profession, the more this factor can cause stress, if not fear. The decision makers might accept mistakes from a novice, but they expect an expert to know *everything*. But you cannot, do not, and will not ever know everything. Prepare, prepare, prepare, but accept, accept, accept that you will make mistakes. Some will be egregious. Some mortifying. One thing you can count on, though: if you do nothing, no mistakes will be made; but neither will any decisions. It is a fact of life. If you lead, you will make mistakes.

No matter your level or reputation, when you make a mistake, there is no reason to fall on your sword and quit presenting. And there is no reason for any mistake to unduly add to the fear equation. Openly admit the mistake, correct it to the extent possible, and go on. It is just that, a mistake, and nothing more. Do not make one mistake a mountain.

Troy Aikman, the former football star, started his quarterback career with the Dallas Cowboys by winning zero games in his first season. He had eleven losses. He stayed at it, though, and three years later led his team to win the Super Bowl. After his football career, he became a favorite sports commentator. He did not let his mistakes take him down.

LeBron James, one of the greatest basketball players ever, started in 600 games. Looking at his statistics, he *missed* 53% of his field goals and 77% of his three-pointers. Even when he had free throws, with no one trying to block or guard him, only fifteen feet from the goal, he *missed* 25% of them. And he is a professional, by anyone's standards a top-tier athlete. You cannot be perfect.

Brett Favre, the longest-running quarterback, an icon of football ability, played in 300 games during his career. He threw 10,000 passes. He *failed* to complete them 38% of the time.

And all these players represent the top level of performance. The *top*. You cannot be perfect. You will make mistakes. How you handle those mistakes will tell your audience more about you than ten things you do perfectly. Establish a reputation for integrity and honesty. Let the mistakes sharpen the blade, not dull it. Examine the mistakes, understand why you made them, and then correct them. Correct them, but do not dwell on them. Do not let mistakes become part of your fear factor.

Focus on your successes, not on your failures. Focus on your wins, not on your losses.

GENETIC EXCUSES

A fourth major reason for wizard stage fright is natural introversion, which may be the reason you went into technology or science in the first place. You may not be able to change your genetic disposition, but do not use genetics as an excuse to have fear. To some extent, everyone has this fear, but professionals learn both to hide it and to channel it into productive externals. You, and everyone else, might be born with this fear, but neither you nor anyone else need be subjugated by it.

Whether you have irrational fear or normal fear, take every opportunity you can to make technical presentations, even if to your own small group of peers, even if informally. This exercise will reduce the unreasonable fear and give you practice in channeling the energy. You may simply have to work harder at this than others more genetically disposed.

THE UNFAMILIAR

Any unusual or different situation exacerbates the fear of public speaking. Maybe the audience is the largest you have ever experienced. Maybe you presented many times to managers, but never to the president. This transport from your comfort zone to the Twilight Zone can increase your fear, big time.

A large audience is intimidating. You may be accustomed to presenting to 5-10 people, but now you are asked to present to 500. I have found that large audiences are much easier than small audiences. With a large audience, there are generally few interruptions, much broader acceptance, and it is much easier to get the audience on your side. Humor, very effective in any presentation, is far easier with a large audience than with a small one. In a large audience also, the questions are usually more basic, broader, and simpler than in smaller groups. Unfortunately, few decisions are made in large meetings. The presentation leans much more to public speaking and significantly away from persuasion.

Unfamiliar equipment can set off the fear factor. Simple things like a lapel microphone, odd acoustics, an unfamiliar auditorium, incorrect attire: all of these are trigger mechanisms. Of these, more in Chapters Eleven and Twelve. Suffice it now to say that you should eliminate as many of these distractions as possible, before the meeting, so that they are not even a factor in the actual presentation.

LACK OF CONTROL

Another factor that builds fear is the lack of control. Wizard technologists can be control freaks. As you grow technically, you naturally take on more responsibility within the organization. As you mature in your technology, your responsibilities and the size and number of your technology teams will naturally grow as well. You will have more people acquiring and interpreting data, more people on whom you must rely. It will no longer be first-hand involvement. You are prepared in the meeting, of course, but not to the same level as when you once reported on only your own individual work. You are not in control of some key elements.

In this maturing role, you will be blamed for mistakes made by other people on your team. This happens, but never embarrass members of your team by publicly pointing the finger at them. It is part of your job to be responsible. If you are the presenter, you are responsible for what is presented. Take the blame and go on. Later, you can meet with the team and put guidelines and practices in place to eliminate such errors. After the presentation, or before the presentation, you can counsel them, tutor them, even fire them. But never, never, never blame anyone else for mistakes that occur in your presentation.

I hear presenters who, during their presentation, see minor typos in the printing of their presentation or in the viewgraphs they are showing. I am disappointed to hear these high-level technologists blaming an administrative assistant or an underling. If you present the material, it is yours. Accept responsibility. Be bountiful with praise, but a miser with blame. When it comes to mistakes, own all of them.

This lack of direct control generates fear in a technologist. Teams can produce massive feats that can never be accomplished by individuals, but there is cost. They can also produce mistakes for which you, the technology leader, must take the blame. Effective teamwork is something that comes with experience. The only way to overcome this fear from lack of control is to develop a trustworthy team and have adequate checks and balances. We discuss the checks and balances, later, of course.

Assume you are correctly prepared. You know the material. You have mastered all the skills we discuss in this book. You are still unduly nervous. What to do?

Two things. First, channel the undesired energy into effective movements, articulations, choreography, and inflections. If you do this, almost all the stage fright will be reduced or disappear within a minute or two into your presentation. All you need do, then, is the second part. Squelch the undesirable external signs that reveal to an audience that you are nervous. You may still be nervous, but no one will know it. In time, it will go away entirely. Here is how to do that.

HARMONIZE WITH THE ENVIRONMENT

Avoid obvious factors that can add to nervousness. Dress like the audience. If the decision makers all wear coats and ties, or jackets and heels, then you should dress accordingly. Do not waste energy or assert independence with a fashion statement. This will do nothing except to reinforce a stereotype on your part, and produce intellectual distance between you and the decision maker. It adds to your fear.

Be prepared for hot and cold flashes. On occasion, you may be sitting idle a long time before you actually speak. During this time, you are not really "on the spot," so to speak, so you relax. If you become chilled, it accentuates an apparent nervousness—folded arms, crossed legs, shaking or bouncing. On the other hand, if you are comfortable, you feel much more confident and not as nervous. I find most conference rooms quite cool. Thus, a jacket or sweater, when appropriate, is a great solution for those times when you may be waiting a long time. Hot flashes come and go. Unbutton a coat or jacket and just wait a few minutes. No one will notice.

NOT FINISHED YET

No engineer ever felt the joy of thinking he had enough data. There is always something else to check, something else to improve, more to verify. To the wizard mentality, the data are always incomplete. What you must understand, as a wizard, is that no decision maker ever made a decision having *all* the data, either. If a decision maker waited until all the data were in, she could never beat the competition. She could never get the product to market in time. Her decision making skill is built around being able to take what data she has and make a strategically profitable business decision.

To the wizard, there are always more ideas to examine, of course. So, when a presentation is requested, fear arises because more data is needed in order to be 100% confident. Forget being 100% confident. It only happens in Hollywood. You will be required to make every presentation long before you are 100% confident. Worry less about whether you have *enough* data and worry more about whether it is the *right* data for the decision makers.

There is no solution for this, just mitigation. All you can do is quantify the data you have and make predictions as best you can based on the facts you do have. The best way to mitigate this fear is honesty and integrity. The worst mistake is to think that you can do or say nothing until everything is complete. It will never be complete. The best wizards know when it is acceptable to make predictions and then they quantify their confidence. Above all, maintain integrity and develop a sense of trust with the decision makers.

TRUST

Recall earlier our mentioning Kennedy's decision at the "Bay of Pigs"? Kennedy, with no experience in photo interpretation, was clueless as to what was in the photographs. No amount of repetition or additional technical details would have made any difference. Kennedy was incredulous whether the Soviets would be so bold as to put missiles right there under his nose. How was it that a photo interpreter, a technologist, was able to convince a president? People who were in the meeting said Kennedy based his decision on *trust*. Kennedy believed that the technologists were trustworthy. He trusted what they were saying.

Eliminate fear by building a reputation of *trust*. A technologist must be trusted by the decision makers. Having that trust goes a long way towards eliminating any fear of presenting. Developing trust is a difficult problem for the new presenter because how does one develop it? Part of the answer lies in this book. Putting the material together in an understandable way and knowing how to speak to the mind of the decision maker: those are significant components. They will go a long way towards developing and establishing trust. Over time, trust will be developed, but it should caution the technical presenter to ensure that every presentation is technically sound and every presentation soundly presented. Trust, once compromised, is rarely retrievable.

I was salutatorian at my high school. I had to make a speech at graduation. I chose *Integrity* as my topic. I thought it was important then and I know it is even more important now. I explained that an integer is a whole number, it is not divided. So it is with integrity. A person with integrity is not divided. He does not say one thing and do another. He does not think one thing and act contrarily. I told them at graduation that it is this unity, this lack of division, that gives a person with integrity the inner strength to do more, to accomplish more, and to be more.

From an engineering and scientific viewpoint, you cannot always be correct. No one can be and most decisions must be made before all the data is in. So, you may not be correct all of the time, but you must have integrity 100% of the time. That means being able to recognize and acknowledge early on when you have made a mistake. Success and failure: both are in the trust equation. Instead of letting the lack of data produce fear, let it produce integrity.

TURN OFF THE BROADCAST

Let me tell you what fear is. A few months ago, I decided to solo skydive. I had never done this before and have not the faintest idea what possessed me to do it at this point. I spent six hours in the training class and then pulled on a helmet and squeezed into a jumpsuit. I strapped the goggles on and wrapped the altimeter on my wrist. I fastened on the parachute that some $6-a-parachute unknown person had packed. I was going on what is called a flight-assisted jump. Although I would be tethered to no one, two skydivers would jump beside me, just in case. Hmmm.

By the time I stood in the flight line to board the twin engine Otter, my stomach was a Gordian knot, my glands were pouring sweat, and my eyes were ping pong balls. My breathing was short and infrequent. I climbed on board the school-bus-yellow airplane, careful not to run into the side propeller spinning at 1500 rpm. The plane skittered over the grass runway. The ascent was abrupt and steep. The summer heat was stifling, 103° F. No air conditioning. The noise was into 90 decibels. My instructor yelled to me over the roar of the engine, wind, and people, "Look down on the ground and orient yourself to the parachute drop site." I stood crouched against the sweaty bodies packed beside me. I twisted to look out the little window, the plane turned, the fuselage was canted at a 30-degree angle. I was getting dizzy. At 3,000 feet, someone slid open the exit-drop door. The summer wind rushed in like a demon. When the door is open, everyone must unbuckle their seat belts. (Want to guess why? If your parachute canopy should unpin and go flying out the open door, it's better you go with it.)

The noise of the plane, the noise of the wind, the noise of people: it was all building in a crescendo. It took about 15 minutes to reach 13,500 feet. Several skydivers jumped to oblivion. Then it was my turn. I stooped, almost crawled, to the open door that looked out onto the world. I put both feet along the open cavity as I had been instructed. I looked down, as I had not been instructed. Three miles below me was the ground. The wind was beckoning me like the devil to come play. My feet were a quarter-inch from being out the door. My butt was fanning the breeze in a knot. My head was outside the aircraft, praying. My knuckles were white from squeezing the door frame. For the first time in my life I now knew terror.

I went through the motions that I had been taught, "sky," "up," "down," "'arch." I let go, sucked out of the womb into nothingness. I dropped at 100 miles per hour, which is a thousand feet every 6.8 seconds. The plane, my last hope on Earth, disappeared over me in an instant. I went into my practiced arch. The wind was tearing at my cheeks and ripping at my suit. I checked the altimeter. 12,000 feet, 11,000 feet. I reached back three times to make sure I could find the pull chord. 8,000 feet. I looked down for an instant. Dear God. 6,000 feet. I "waved off" so that any skydiver fool enough to be above me would know I was about to "pull." I have not the foggiest notion what that fool might have done about it. I reached back with my right hand and found the rip cord. I pulled it out and let it go. Nothing happened. Oh, God.

The skydivers at my side plummeted to the Earth and disappeared below me. What? Suddenly, I was yanked up like a fish on a hook. I was floating in space. My parachute had opened, after all. I hank God. I was not going to die, maybe. I looked up. The parachute canopy formed a nice rectangle, just like they said it would. I was drifting slowly down to Earth, where I planned to stay for a long time. I took the first breath I had had in five minutes. I raised my arms and gave a Tarzan yell. I was alive, at least for the moment. I cannot describe how it feels to go from absolute terror to peace in such a short time. So, the next time you are nervous in a presentation, consider that there really are things much more scary. And, do not do them.

An audience can detect nervousness in only three ways: an unsure voice, shaking hands, and unnecessary body movements. Secondary effects like sweaty palms, dry lips, and knocking knees are annoyances to you, but scarcely perceivable to an audience.

Let us start with the secondary effects. Dry lips and sweaty hands are common. No one will notice them. If you need to shake hands, wipe your hands surreptitiously on your pants or skirt. It may not be the essence of etiquette, but it will leave you with a nice, dry palm. Put some lip balm on before the presentation.

Undesirable and uncontrolled movements like pacing and moving about must be channeled into controlled articulations. Effective articulations and choreographed movements are covered later. Annoying habits like shuffling papers, rubbing the face, scratching the nose, wetting the lips, and similar, are just that—habits arising from nervousness. Practice not doing them. They are not so strongly coupled to nervousness that they cannot be overcome mostly by consciously *not* doing them.

Those shaking knees are not noticeable with slacks or pants. It is noticeable to the owner of the knees, not to the audience. If you are sitting down before the presentation, it is okay to bounce them up and down *gently*. Many people do that normally. Just do not be conspicuous about it. When standing, one way to mitigate the perceived effect is to stand near the table or near something that blocks the lower part of the body, at least until your presentation gets started. I do not recommend standing behind a massive podium except under the most formal of circumstances. Some movement is quite beneficial to the presentation so maybe just moving to a different position every five minutes or so will help. If this is really a persistent problem, I definitely recommend not wearing shorter skirts as this may bring attention to the very area you do not want noticed, especially if the platform is raised.

Hands shaking? Just leave them hanging at your side or rest them occasionally on a platform or podium at your side (not in front of you). Mostly it is a matter of what not to do. Do not hold a pointer as this will leverage the movement to be conspicuous. Do not hold papers as they bring attention. Do not hold the hands where they can be seen shaking. At the side is a good place because they go into the folds of the slacks, pants, or skirt and break the effect. They will quit shaking when you give them a specific place to be and a specific function to perform, like nothing.

The only truly problematic evidence of stage fright is the voice. If the voice quivers, this is noticeable. I knew a woman who was a high school drama teacher. Whenever she got in front of an audience, her voice wavered and faltered continuously. (I never understood how she could teach drama.) If this happens, project more forcefully. Quiet, timid speaking accentuates the situation, whereas pushing the volume through and projecting will help mitigate it. "Be bold and mighty forces will come to your aid." Shout it out if need be and the wavering will stop.

Smiling is always a good thing. It changes your voice. Smiling changes your attitude, too. Do it often. The first thing I do is walk to the podium, smile, and say nothing for a moment. It puts you in control of yourself.

Do not think about being nervous. Just give the nervous body parts something to do and soon you will forget all about them. Do not think about nervousness, do not notice it, do not give license to it. To get it out of your mind ahead of time, begin to rehearse the opening lines of your presentation. Think about the decision makers reacting favorably to your every word. Think about your past successes. (You should have forgotten any failures, already.) You are in control. Force negative thoughts away by filling your thinking with good, solid, positive thoughts.

Think good thoughts throughout the presentation and not negative ones. This is not the time for self-critiquing. It is a time for self-praise. Animadvert after the meeting, not during.

SLACKEN THE PACE

A person with stage fright, one who exhibits nervousness, hurries through the presentation. In a subconscious, or even conscious, attempt to "get off the stage," the pace increases until it is noticeably too fast. It sends out a warning signal, "I'm really nervous and need to finish this as fast as I can and get back to my real life." The presenter may not even notice how the pace has quickened, but all effects of intonation, emphasis, phrasing, and articulation vanish and haste takes over.

Slow down.

Only about 5% of novice presenters speak *too* slowly and even then it is not the pace, but the "and uhs" that make it seem slower than it is. Take it as a general rule that you need to slow down your rate of delivery. This will help ease your own nervousness and certainly do wonders towards transmitting a sense of confidence.

If a manager or decision maker gives you the sign to "speed up," it is not a request to vocalize more rapidly. It is a request to be more succinct and give less detail. It is a request to leave some things out, get on with the presentation, or quit being side-tracked by every notion.

Even if you do all these things, in the end, you should still be slightly nervous, still have a little stage fright. Why? Stage fright is the adrenaline that helps you hone your game. Remember? Besides, if it were easy, everyone would be a good presenter. You are a good presenter because you know how to conquer your fear.

Once you conquer this fear, you appear to the audience as a confident presenter. When you appear to be confident, this goes a long way towards removing the wizard stereotype. You are not just a techie who should be in the lab soldering wires, you have an accomplished presence, even when before decision makers.

So, you develop the confidence, or at least you show external confidence, but does that eliminate the stereotype? No. It does not. It is just a start. Now you must add something to that confident appearance. You must have a presentation worthy of the decision makers and deliver it in a manner that is persuasive.

Mitigating the Stereotype

- Learn to exhibit confidence and assuredness.
- Fear of public speaking is common to everyone, but it can be an engine of great power. Use it to your advantage to produce a better, more persuasive argument.
- Identify and channel the external signs that transmit a lack of confidence.
- Focus on your end goal, not your fears.
- Be adequately and correctly prepared.
- Know how to handle another expert who would bring controversy.
- Take responsibility. Never, never blame others for a mistake in your presentation.
- Have integrity. Develop a reputation of trust.
- Control every element you can. Reduce the surprises.
- Slow down. Never try to gain time by speaking more rapidly.

Chapter 4: What, What, What
Forsaking Standard Approaches

It doesn't matter how beautiful your theory is, it doesn't matter how smart you are. If it doesn't agree with experiment, it's wrong.
— Richard P. Feynman

*M*ost colleges encourage engineers to take courses in communications and public speaking. In these courses, would-be technologists learn the basics. The instructors strive to reduce the "uhs" and the "and uhs" and to eliminate annoying actions like pacing, fig-leaving, and chewing gum. There is usually time for students to practice in front of the class. Teachers and fellow classmates critique and give feedback on the presenter and on the presentation's organization, delivery, clarity, and effectiveness.

Yet if we expect to transition from influencing instructors and peers to influencing decision makers, we must increase our skills beyond these rudimentary essentials and build to a higher level of persuasive capability. How, and in what ways?

The first requirement is to develop a strategy for the presentation itself. What basic structure do we build upon? Where do we start, where are we going, and how do we get there?

The schema of most engineering-communication courses follows the general rules of public speaking. This is a mismatch. This approach ingrains a traditional approach, but that approach is ineffective, even counterproductive, for technologists persuading decision makers. What is the traditional strategy and why is it so ineffective in persuading decision makers?

THE REIGN OF THE *THREE WHATS*

Think of this scenario. You are a new engineer at the company and you have been asked to make a presentation to the vice president of the organization. You ask senior engineers, "What outline, strategy or general approach should I use in the presentation?"

The answer, invariably, will be, "Use the *Three Whats*, tell them what you plan to say, tell them what you say, tell them what you said. When you present technology to any decision maker, use the *Three Whats*. The *Three Whats* is a surefire proven strategy."

Proven, yes, but not for the audience we target, not for decision makers, and not for technology persuasion.

Tell them what you plan to tell them, tell them what you tell them, then tell them what you told them. Why is this supposedly time-tested strategy ineffective for decision makers?

First, let us analyze when it is effective, and why. Then we can show why it is not effective, any time, for decision makers and why you must never use it.

The staying power of the *Three Whats* starts with the familiar and effective use of triplets. For example,

Father, Son, and Holy Ghost;

Faith, Hope, and Charity;

The Good, the Bad, and the Ugly;

Going, Going, Gone;

and Red, White, and Blue.

The list goes on. Every good journalist knows the effectiveness of using three conjunctives to bring home a point. The three elements may be different in many ways, but they are grouped together because they are alike in at least one important way. The effectiveness can be enhanced by other grammatical techniques such as alliteration or rhyming.

The use of three is effective in other ways. Three-fold repetition is a common memory aid. I turn on my radio while driving to work. The sports talk show comes on. It breaks to a commercial.

Are you tired of your spouse? Do you need a change in life? Do you need a lawyer to help you in this difficult time? Then you need to call Ronald and Ronald Associates and get the real professionals. That's Ronald and Ronald Associates. Their number is 284-903-6622. That number again is 284-903-6622. Again, that number is ...

I reach out quickly and punch the channel button, hoping to jettison Ronald and Ronald before the phone number lodges in my short term memory. It is annoying. It rings in my head. *284-903-66...* Nevertheless, it must be effective because Ronald and Ronald keep attracting fatigued spouses. Many advertisers successfully ply this annoying technique.

Now, we have shown that triplets are resonantly pleasant and a great help in memory. So, since triplets seem to be effective and have years of precedence, why not use the *Three Whats*? The *Three Whats* would seem to be a good way to structure your presentation to decision makers, would it not?

No.

Examine the triplet idea, again. This time, make it a technology concept. Give it three repetitions, then check to see how the audience is doing. Did they remember it? Here, let us try it with one of Maxwell's equations.

$$\text{First time,} \qquad \nabla \times \vec{B} = \mu_0 \varepsilon_0 \frac{\partial \vec{E}}{\partial t}$$

$$\text{Second time,} \quad \nabla \times \vec{B} = \mu_0 \varepsilon_0 \frac{\partial \vec{E}}{\partial t}$$

$$\text{Third time,} \qquad \nabla \times \vec{B} = \mu_0 \varepsilon_0 \frac{\partial \vec{E}}{\partial t}$$

Now, three times, do you have it? "What?" you say. "Do I have it? I don't even know what it means. And, I don't know why I would even care what it means." Hmm. Now, that is a problem. Here we have repeated it three times, but you are no closer to understanding, no nearer to comprehension, and seem to have no motivation to learn it.

Perhaps that was extreme. We should try something simpler. I will begin by explaining how one finds the roots to a quadratic equation. Here, start with the equation:

First time, $3x^2 - 5x - 4 = 0$

Second time, $3x^2 - 5x - 4 = 0$

Third time, $3x^2 - 5x - 4 = 0$

Was that better? "No," you say. "It doesn't make it any simpler just because you repeated it. If I did not understand it the first time, I won't understand it the second or third time, either. If I do not know why I need it the first time, I won't know the second, either."

Now, we are getting to the heart of the matter.

Most technology wizards would not consider this last example as a complex problem. But the point is that without some knowledge of algebra, repeating this equation a hundred times will not make it understandable, much less solvable.

Ask any science teacher how many times, and in how many different ways, she must describe Newton's Laws before the students understand those laws and can apply them. It is much more than three times and Newton's Laws, albeit profound, are not difficult to explain or understand.

These are exemplary illustrations of how difficult it can be to explain technical material to a non-technical audience. Now, add the fact that you want this non-technical audience not only to understand the information, but to act on it in a preferred way, and you have the problem of the wizard who is facing decision makers.

Do not get me wrong. The *Three Whats* is effective for a different audience and a different type of presentation. An example will illustrate the *Three Whats'* effectiveness when given the right environment.

Having experienced the annoyance of lifelong poor vision, I jumped on the bandwagon for corrective surgery. I chose a doctor who had performed this surgery thousands of times. I attended his seminar on *Frequently Asked Questions*. When I arrived there, there were about twenty in attendance. Drinks and refreshments were available. The doctor began his presentation with, "I'm going to describe a surgical procedure that will likely restore your vision to almost twenty-twenty." This was the *First What*, the "*What* I Plan to Say."

Then, he described the procedure and its benefits, the "*What* I Say" part. This was the data, the facts, statistics of the number of successful surgeries, expected results, cost, and time required off the job.

Finally, he wrapped up with a summary of "Here's *What* I Said."

It was the standard *Three Whats* scenario and it was very effective. I signed up.

But what made it effective in this particular situation?

The *Three Whats* is appropriate and effective when certain criteria are satisfied. And, all the criteria must be satisfied or the *Three Whats* is ineffective.

First, the presenter must be acknowledged by the audience as an authority on the subject, that is, the speaker must be recognized as an expert expounding information to an audience that acknowledges his expertise as superior and their's inferior, at least on that subject.

In the example, the doctor had a friendly audience who did not question his knowledge and who admitted only limited knowledge of the subject themselves. But, they were vastly interested in what he had to say, eager to have better vision, and, by all means, wanted to enable the outcome. They all knew why they were there. The audience had no reason to animadvert either the authority of the speaker or the effectiveness of the procedure. It was what it was. The audience was just inquiring, *voir dire,* not judging.

This doctor, then, had a special audience and a special situation. Not a single person in the audience was the doctor's CEO, nor the president of his company, nor a supervisor, nor selected experts critiquing his surgical procedure. It was not the FDA or any other governing body that might exhibit contrariwise opinions. It was not a group of venture capitalists contemplating investment. Such audiences would be anathema to the use of the *Three Whats*. Why?

To begin with, decision makers do not capitulate to technical expertise, even if it is acknowledged. Their position, title, and energized egos usually makes them feel superior, your expertise notwithstanding. After all, *they*, not you, are making the decision. They, not you, deal the cards. In any case, the wizard's technical knowledge, regardless of its depth and breadth, is not acknowledged as superior to their business and decision making knowledge. Second, decision makers may or may not be receptive, *a priori*. They may not even know why they should be interested or even listen. They are busy people. Sometimes, a good bit of persuading must be done just to convince them they need to hear what is being said. We have shown that the *Three Whats* does nothing to aid in persuasion, itself. It informs. Third, decision makers may discount the information presented, for whatever reason, and make decisions based on other criteria, as we discussed earlier.

Domain of the Three Whats

- Speaker is recognized by the audience as an authority.
- Speaker is providing mainly instruction or information.
- Audience is of a special type, namely:
- Acknowledging—recognizes and respects the speaker as an authority on the entire breadth of the presentation.
- Receptive—requires or needs the information for fulfillment and is favorably disposed to the information, *a priori.*
- Enabling—is capable of and desirous to enact the recommendations of the speaker, *a priori.*

We showed that the *Three Whats* is effective for the right audience. We defined the *right* audience for the Three Whats. What remains is to illustrate why it is so ineffective for an audience of decision makers.

Let us try the *Three Whats* on two real decision makers, John and Suzanne. Their decision is one of marriage: should they marry each other? Convince them to make that decision. You, the wizard, are the presenter.

Armed with your *Three Whats* strategy (how can it fail?) you start the presentation. You look up at the clock and check your time. It is seven o'clock in the evening. The organist begins a processional. It appears they are starting the wedding.

What? Whoa! You stop the music, ask everyone, including John and Suzanne, to come immediately into the auditorium and take their seats and be seated with their respective entourages. They were ready to march in, but the *Three Whats* protocol must be followed. You need to get them all on the same page, so to speak, and tell them what you plan to tell them. Suzanne comes in with her bridesmaids and John with his groomsmen. They all look a little bewildered, but being a typical wizard, you either do not notice or, noticing, you ignore it completely. You look at this astonished group and the audience and deliver your *First What* with absolute and full confidence.

"Dearly beloved, parents, relatives, and friends, I want to tell you about the ceremony I plan to perform. First, we will have an organ prelude. I'm sorry I let it start before I gave the introduction. The groom will go out from here and enter again with his groomsmen. Then, the bride's attendants will enter, one after another, walking slowly to a traditional march that goes something like this," whereby you hum a few measures of Pachelbel's *Canon in D* so that the audience will be sure and know it when they hear it. (Remember, tell them what you plan to tell them.)

Now, if you are like most engineers and technology wizards, you will get sidetracked at this point and feel compelled to tell them a little known fact that the bridal march is a score from Richard Wagner's 1848 opera, *Lohengrin*. You show your knowledge and embellish that Wagner, albeit a great musician, was anti-Semitic. (Are there any Jews in the audience? Well, never mind.) Add that the organist will play it in the key of F major at an *adagio* tempo of 76 beats per minute. A true-dyed-blue wizard will even describe the pipe organ, who made it, how old it is, how many stops it has, the size of the pipes, how sound is produced in a closed pipe, maybe even throw in a few physics equation and discuss "lambda" and "nu" and the velocity of sound at ambient temperature. All this information, albeit technically accurate, is entirely irrelevant to John and Suzanne's decision, and to everyone else in the audience. However, since it is correct technical information, as an engineering wizard, you feel you must include it. The fidgeting audience and Suzanne's once beautiful, but now dagger-eyes are transparent to you. You know deep down inside, being a wizard among wizards, that they all want and need this additional information—because you know it and feel innately compelled to show you know it.

Where are we? You are still in the *First What*. Much time has passed. Next, you tell them that the bride's friend, Melinda (whom the bride has known since the fourth grade, with whom she was friends in high school, who owns a red Mustang like yours, and so forth), will sing Schubert's *Ave Maria*. (We will not go through all that again.) You continue with your now interminably long *First What* saying, "Following Melinda's song, I plan to query John and Suzanne, in turn, as to their nuptial commitments. To save time, I intend to limit their choices to just each other. I also plan to ask if anyone objects, so if you do object now is the time to think about how you would word your objection and maybe write it down. Assuming no one objects and assuming John and Suzanne both agree to marry only each other and not anyone else in the room, ha-ha, (wizards love wizard-humor) then, I will declare it so and give a brief statement of my credentials. We'll end with a kiss and I expect John and Suzanne will hurry out the front door. We have other exits in case there is an emergency and they, or anyone, need to leave sooner. The exits are in the back and there are two more on the sides, here, near the front."

"That will be my presentation. We should be done in about 20 minutes (you have already taken more) unless you have questions and please stop me if you do have questions, but for now we will start the ceremony."

The *First What* is finished. You told them *What you plan to tell them*. You took too long, though, and a good portion of your planned time is eaten up already. The caterers for the reception are on a strict schedule. Time is short. You decide to make up for it by hurrying through the *What I Say* part and maybe have only John take the detailed vows, just to save time. Suzanne looks pretty committed already.

You begin the *Second What*, the ceremony itself. In the *Second What*, the organist plays, Melinda actually sings the song, and everyone else does their part. You quickly shuffle through the "if anyone has an objection" part because there are no contingency plans if that occurs and time is short, anyhow. You skip Suzanne's vows. You make your pronouncement, they kiss, the finale starts, they rush down the aisle, music is playing, the audience is standing, everyone is smiling, and you yell ...

"Stop, everyone! Please. Stop the music! Stop!" You must shout above the din. "Come back and sit down, please, please, everyone! Sit down. This is important, please."

It's time for the *Third What*. You must tell everyone what you just said. It is easy for them to forget and most people have to hear it three times to understand it, anyway. At least that's the doctrine of the *Three Whats*. So you begin the *Third What*.

"Dearly beloved, I now need to give you a summary of the ceremony we just performed. John and Suzanne, would you be seated, please? This will only take a few minutes and I will be done. Now, John, you said you would take this woman. And, Suzanne, you did not take the vows because I was short on time and I just skipped it, assuming you would say, 'yes.' Remember, John and Suzanne? That's what we did. Suzanne, your friend, Melinda, sang a song about eternal love. It was sung in the key of G." (You forgot to tell them about the key of G in the *First What* so you add it now because it is a fact.) "So, now that everyone is sure of what they saw and heard, unless there are questions, that's the end of the ceremony."

Huh? Does this ridiculous application of the *Three Whats* make sense for these two decision makers, John and Suzanne? No, and it doesn't make sense for any other decision makers, either.

Perhaps another example will help. My octogenarian father tells jokes using the *Three Whats* approach. Is this the best structure? Let us see.

First What: He says, "I want to tell you a funny joke. It's about this couple that have been married forty years and the angel Gabriel wants to reward their faithfulness by granting them each a wish. The punch line is about being married to a younger woman, but not in the way the husband thought. Have you heard the joke?"

I say, "No, I haven't heard it." (Actually, he has told it to me several times.)

Second What. "Well, there was this couple that had been married 40 years, each one sixty years old, and the angel Gabriel decides to give them one wish each for having been faithfully married for so long.

The woman says, 'I've never been on a cruise. I want to take a cruise.' Angel Gabriel says, 'Poof. No problem.' And off she goes on a cruise.

Angel Gabriel asks the man for his wish. 'I want to be married to a woman 20 years younger than me.' 'No problem,' responds Gabriel, and, poof, turns him into an 80-year old man." The *Second What* is finished.

"Did you get it?" my dad asks, possibly because I'm not laughing.

Now, the *Third What* comes. "Maybe you didn't get it," he bemoans. "The man thought he would get a new wife 20 years younger, but Gabriel made him 20 years older than his current wife." I smile, tell him I get it, and end the presentation before my dad goes to his backup material.

Now, one might object that the joke is not an example of decision making, but it is a very good example. What is the decision? The decision is whether or not to laugh. You must *persuade* the audience to laugh. It must be a spontaneous decision, though, and that is precisely the point. The teller of the joke wants a predetermined outcome—laughter. He wants the audience to come to its own conclusion automatically, based on the effectiveness of the presentation. The joke teller wants to *persuade*. The same for the wizard making recommendations to a decision maker. The wizard does not want laughter, of course, but the wizard wants a predetermined, spontaneous response—a decision in his favor, an approval of his recommendations. For the very same reasons that the *Three Whats* fails in a joke, it fails to persuade decision makers in a technology context.

There are many other examples. Sales and marketing pitches would fail miserably if they used the *Three Whats*. Think of any television commercial, reinstate it into the *Three Whats* and then apologize for how more painful it would be. By now you should recognize that there are times when the *Three Whats* is a woefully inadequate, often absurd, approach. Why is this? Why do the *Three Whats* not work for decision makers?

First of all, senior leaders, decision makers, neither want nor need to hear something three times. Granted, approval for your request may require six presentations, but that's not because they did not understand it the first time. They did understand. It is because they either did not accept what you said, felt there was insufficient information, were not ready to act, or simply had other, more pressing matters at the time. Ideas, programs, and projects are hard to sell, even very good ones. But don't torture and handicap a good idea by using the *Three Whats*.

A second reason the *Three Whats* does not work for decision makers is because presentations to decision makers are (or should be) much more than communicating status or giving instructions, which is the primary domain of the *Three Whats*. We use the term "decision maker" for a reason. If the goal of your presentation is to simply to give status, then send a report or an email and do not waste the time of a decision maker. Again, these are *decision makers*. They are called that for a reason. They receive information, they process information, they make decisions. As a technologist, you are there to present facts, results, conclusions, new ideas, different concepts, or innovations. You are there to persuade, to direct action, to endow them with something worthy of their powers to decide. Remember, when we profiled the decision makers in Chapter Two? Those executives wanted to hear "something they did not know already." They do not have the time nor inclination for useless repetition, and the *Three Whats* is just that.

A third reason the *Three Whats* does not work is because decision makers think too quickly and almost immediately get out of phase with your presentation. They are there to make decisions—they start making them, whether you are ready or not. If you use the *Three Whats*, they will logically take a position and form an opinion during the *First What* and then you will spend the rest of the presentation (if you get to finish) trying to dislocate the opinion you forced them to form during the *First What*. While you're detailing the *First What* and exhausting their attention span, they will flip to the back of the written material you gave them to see the conclusion, form an opinion, and be waiting for you to get there. You might try to countermand this by letting them see only one page of the presentation at a time, but this is not good either, as we will discuss later. And, regardless, it will not keep them from forming opinions.

A fourth reason not to use the *Three Whats* was illustrated poignantly in the example of the joke. To be funny, the punch line must be poignantly clever, insightful, or compelling in order to stimulate a favorable reaction from the listener. You can hardly stimulate a listener if you tell them what the joke is before you start. Similarly, a technology presentation needs logic that builds and then gets to the point quickly, like a supersaturated solution suddenly crystallizing. The technology presentation may not have a funny punch line, but it does need punch. Setting it up ahead of time (*First What*) and then repeating it (*Third What*) takes the punch out of it. The *Three Whats* lets the air out of the tires. The vehicle goes nowhere. It ensures the presentation will be flat and most technology presentations already have a problem with this.

When presenting to decision makers, forget the *Three Whats*.

Are there other options for structuring a presentation, other than the *Three Whats*? Of course.

I attended a dinner with Condeleezza Rice, former US Secretary of State. Afterwards, she was asked to speak regarding any subject she chose. She stood and proceeded, without notes, to deliver a riveting forty-five minute "presentation." What was her structure? It certainly was not "here's what I plan to say, here's what I say, here's what I said."

To the audience, the effect was stream-of-consciousness, not knowing what to expect next, constantly in anticipation, eager to hear each thought and its ramifications. And, of course, if you have ever met her, her locution is precise, her thoughts surgical, her poise captivating, and her presence riveting.

Internal to the speaker herself, though, was a well-laid plan of seven major topics. First topic: she started with a brief description of her life since leaving the White House. "I now have the luxury of calmly reading the newspaper each morning murmuring, 'That's interesting.'"

She introduced each subject with a rhetorical question or a statement.

Second topic: "Defending the United States is still of paramount importance." She then proceeded to talk about terrorism, what had been done, what should be done, its long term outcome, about Afghanistan, Iraq, Iran, and specifics about governments and dictators. She talked about democracy and the founding fathers.

Why the Three Whats is Ineffective With Decision Makers

- o Its strength is redundancy—anathema to quick thinkers and busy decision makers.
- o Requires unnecessary, additional time from decision makers who already are short-selling that commodity.
- o Flies in the face of the decision maker's need to hear "something she does not know." Only one *What* of the *Three Whats* is original; the other two are repetition, even if said with different words and in different ways.
- o It fosters premature formation of opinions.
- o Lines of demarcation begin to form during the *First What,* before any actual data is given.
  - Opinions form prematurely and incorrectly based on inferences to the *Second What.*
  - Once formed, extremely difficult to get decision makers to change an opinion.
  - Takes the "punch" out of recommendations and leaves the audience flat.

Third topic: "Every home, every lost job is a tragedy." She then discussed the economy, free trade, problems she had in office, and her idea of what should be done.

Fourth topic: "Do I think China will surpass the US as the world leader?" She brought in ideas of censorship, economy, growth, and labor unions.

And so on with all seven topics. She finished with an admonition that ordinary people in the right environment can do extraordinary things, alluding to the title of her book.

So, she had a well-organized, well-planned structure. It was seven unrelated topics that were of general interest. There was no introduction at the beginning or summary at the end. It was spell-binding.

Another presentation for comparison was given in Bangalore, India by Dr. Krishnaswamy Kasturirangan, then Chairman of the India Space Research Organization (ISRO). We were the guests of honor at a dinner he hosted. (It is not difficult to be a guest of honor in India, because the warmth, kindness, and hospitality of these people is unsurpassed.) After the dinner, Dr. Kasturirangan gave each of us a small gift. Mine was a set of carved elephant book ends that sits in my study and daily reminds me of my friends there. He made a short speech. The *Three Whats* were nonexistent.

He talked about his organization and the Indian people who had made the ISRO technology world-recognized. He spoke of current launch vehicles, satellite sensors, and business relationships. He talked about the honor of having us there to celebrate a milestone with them. He discussed future plans and ideas.

The structure of his presentation was past, present, and future: where we have been, where we are, and where we plan to be. It was perfect for the occasion. There was no summary to remind you of what he had just told you. (How could you forget?)

In a meeting in Tokyo with Mitsubishi, an engineer used a different approach. He did not have an introduction at all. There was a title to his presentation so he assumed everyone knew what he planned to present. He got up and just started. I had no problem with it at all, but at that time, I was a physicist listening to a peer. The other decision makers with me drifted into a deep coma. His structure was to have no structure at all, just a single "here's what I have to say." It was ineffective.

So, now we realize there are countless ways to organize a presentation. Different strokes, different folks. But which approach is best for a technologist persuading decision makers? Is there a preferred methodology that ensures success? Yes, and it is the subject of the next chapter. But, do not forget the major point of this chapter and do not use the *Three Whats*.

Organizing a Technology Presentation

o Different audiences require different strategies.

o Target your specific audience of decision makers.

o Do not use *What, What, What* with decision makers.

Chapter 5: Why, What, How
Using the Paradigm for Persuasion

When you do the common things in life in an uncommon way, you will command the attention of the world.
— George Washington Carver

*F*rom the twelfth floor, the view is spectacular, the Pacific Ocean in the distance and downtown Los Angeles honking and fuming below. On the ground floor, the Technology Center is dense with wizards of all venue setting up the latest equipment. It is the quarterly meeting for our Fortune 100 company. In the auditorium the reporters, financial analysts, investors, and business pundits are assembling to hear the CEO give corporate prognostications and deliver the performance numbers for our company's quarterly report.

For this occasion, we refurbished our Technology Center (a big empty room with enough cabling to link together two continents) to make it look like a trade show. We have red carpet, curtains, videos, flashing lights, spinning wheels, specially-equipped vehicles, intelligence-communications equipment, and satellite command-and-control stations. It is a wizard's paradise, although all the wizards are awkwardly attired in heels, skirts, coats, and ties. The top wizards were flown in from across the country to explain everything to the decision makers at the quarterly meeting. More wizards are standing by, ready for any contingency or breakdown. Backups to the backups are available with a single text message or cell call. Bartenders shine their glasses at the little carts that serve as open bars. Tables are full of expensive *hors d'oeuvres*, on real glassware. (Not the plastic imitations we employees get.) Myself? I have a demonstration set up on how we do satellite imaging. We tested the demonstration, validated it, and were ready hours ago. I decide to wander into the auditorium to see how things are going with the business presentations, maybe to be available to answer any technology questions should they arise. (They will not.)

Several hundred decision makers are seated and many more are listening on the teleconference phone at the main platform. The decision makers are mostly reporters and investors—reporters making a decision what to report and investors making a decision whether to invest. The CEO is starting the introduction to it all. He envisions a wondrous future for our company. A couple of the division presidents speak next about their whiz-bang technologies. I can sense the audience getting bored. They want to move on. No one is taking notes. Most are reading emails and sending text messages on their wireless handheld devices.

Now, the chief financial officer (CFO) steps to the microphone for his presentation of the quarterly performance numbers. The decision makers come to attention, sit up in their seats, lean forward, scribble notes, type information into laptops, fervently transmit the information on their wireless devices, and raise their hands to ask questions.

As soon as the CFO sits down, the audience evaporates like rain on the Serengeti. The CEO comes to the platform to end the meeting. I hustle back to the Technology Center to await the expected surge of reviewing decision makers. After two hours, less than thirty-five analysts have journeyed beyond the drinks and the *hors d'oeuvres* and into our demonstration area. Why? They are decision makers. They can only live a few minutes in a technology-rich environment before they start sucking air. The libations encourage them but the attraction is short-lived. They are interested in the bottom line, the financials, not technology words and pictures, things they do not understand to any depth. I manage to get three small groups interested in my demonstration. They ask polite questions, none of them probing. You can spiel technology all you like, but their decision attaches to the bottom line.

Decision makers have their own interests and their own thought processes. If you plan to get their approval in a stand-up technology presentation, then you must target those interests and structure those processes. To do that, you need an effective structure upon which to build your presentation. Of one thing we can be certain, it must be far superior to the effete *Three Whats* structure. Before we describe the correct approach, let us ensure we understand what is wrong, specifically with the *Three Whats*.

Telling decision makers "*What* you are going to tell them" is akin to telling them. It encourages them to form opinions based on the preliminary information of the *First What*. It is then difficult to remove these opinions later in the presentation. The more you tell them up front, the more the opinions they form. And you will not have as much time later on in the presentation to change those opinions because you wasted time telling them "*What* you are going to tell them," without really telling them. The knitting begins to unravel because the *First What* leaves all these ends dangling. And the real presentation has not yet started. The *Third What*, "telling them what you told them," is entirely superfluous. If you need to emphasize what you said, then do that while you are saying it, during the main part of the presentation, giving it the punch and drive to resonate, to sink home.

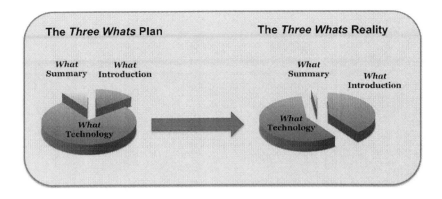

Figure 5-1: The Three Whats is a Losing Paradigm

So, even if you follow your plan and the *Three Whats* is ineffective for decision makers, then what is effective? What is the correct structure for a technical presentation to decision makers? It is this: replace the *Three Whats* with a structure tailored to decision makers.

Instead of using the *Three Whats* structure of "What, What, What," replace it with the strategy, "**Why, What, How.**"

Instead of telling the audience *What You Plan To Say,* followed by *What You Say,* followed by *What You Said,* use instead the structure of telling them **Why** you are presenting (why they are here), **What** you are presenting (the same as the *Second What* of the *Three Whats*), and **How** they can enact the **What** of your presentation. The new structure is then *Why, What, How,* with the middle **What** being almost identical to the *Second What* of the *Three Whats* approach.

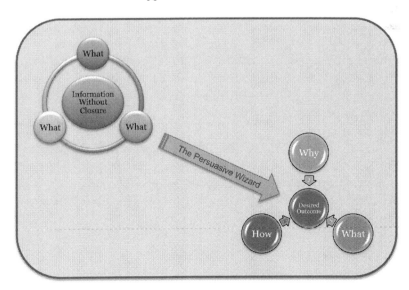

Figure 5-2: Use the Paradigm That Produces Results

WHY

Let us attempt the marriage ceremony with the proper structure for decision makers. You, the technology wizard, are the minister. This time, instead of beginning the ceremony

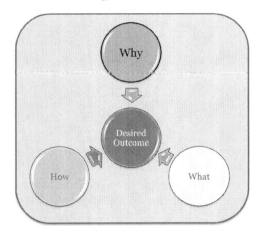

by telling everyone *What* you plan to say, try telling them *Why* they are here, *why* you are here, and *why* you are having this wedding in the first place. Here, let us try it.

"*Dearly beloved, we are gathered together here in the sight of God—and in the face of this company—to join together this man and this woman in holy matrimony.*"

Wow! That sounds effective. Now, everyone knows why they are here, why they made the effort, why they put on a jacket or tie and drove so many miles to this wedding. No waiting to get started—it *is* started. This is worth hearing. It is *worth* it. Instead of going over the mechanics of what you intend to say and all that, you have drilled right to the heart of the matter and nailed it. The angst of an audience squirming to get started is transformed into an energetic expectation of more to come. It is started. You are not just cranking the starter over and over and wishing the whole thing would start. It is started. You have ignited the decision-engine. People sit up in their seats, move closer to the table. They are ready for a decision.

Here is *why* you and I are here, folks—we are here to unite this couple in marriage and they need you to validate and celebrate it. This is the reason we asked you to take time out from your busy schedule to be here and to hear this presentation. This is *why* it is important to you, not just to them, but to all of you. We are here for a single-minded purpose, to unite John and Suzanne in marriage. This meeting and my presentation are, specifically, for John and Suzanne to make a *commitment,* and for you to endorse that commitment.

You establish up front *Why* you are making the presentation, who the decision makers are, and what needs to be decided. You emphasize that this entire meeting is expected to result in a *commitment.* There is a decision to be made. That decision is precisely *why* the decision makers are here, and for no other reason. You posit, at the very beginning, that no other outcome is considered. The presentation to follow will show *What* the commitment is and you will tell them *How* it is to be enacted. But first you establish *Why* they are here. Now the audience and decision makers are ready to listen. Something of interest to them is about to occur. They know that because you have told them *Why* they are here.

Some refer to this *Why* as the "purpose" of the meeting or use a broader term like 'mission'. I dislike both terms. I avoid calling it the *purpose* because the purpose too quickly becomes lofty and remote. The purpose of the ceremony is to get married, of course, but what you really want is *commitment*. Purpose sounds too high and mighty for most meetings, too stilted. Better to think about *Why* you are here and leave purpose to the philosophers. You do not want to begin your ceremony with a discourse on the history and philosophy of monogamous marriage. No. You start with, "We are here to unite."

Thinking of the *Why* as the *purpose* can lead into a historical account of how we all got to this meeting or current situation. This historical voyage is often long, sometimes detrimental, frequently controversial, and brings up old wounds, outdated issues, and ancient baggage. If history and status are important to the presentation, put them into the *What* section. Put history and status in the *Why* section only if it is essential to telling *why* you are having the meeting. The past is past. Rest assured that on this project at least one decision maker in the room will want to forget what is in the past and move on. There is just not much you can do about changing the past and most decision makers are interested in areas where they can make an impact. They look forward, not backward. Start the meeting with *Why* we are here. Do not try to transform it into a lofty purpose.

The term "mission" sounds like a secret government project. Organizations may have missions. Projects may have missions. Meetings do not have missions. If, indeed, you are reporting on some project it may be important that you articulate the mission of that project somewhere in the presentation, but the mission cannot, in and of itself, be *why* you are here. You can argue that they need to hear the mission again, but now you are back into the mode of assuming that by saying the same thing over and over, repeatedly, you will get different results. Start the technical presentation by telling the audience *Why* you are here and *Why* they are here.

Too simple? No. You cannot believe how many meetings I attend when halfway into the presentation a decision maker asks, "Why am I even in this meeting?" What that decision maker means is, "Tell me why you need someone of my rank and position to hear this. *Why* am I here? What is it you are wanting me to do? Here I have sat thirty minutes and heard about technology, but whatever in the world does that have to do with me?" Start the presentation with a forceful *Why*.

The *Why* portion should take no more than about 5-10% of the total presentation time. The *Why* opening is a gatherer, an awakener. It gets everyone on the same page. Not a mission, not a purpose, but a *why* this meeting is worthwhile and *why* you, both the presenter and the audience, are needed.

If the subject is controversial, as indeed many technical-business decisions are, then you may hear an objection right up front, "That's not *why* we are here." If this happens, there is usually a frictional exchange between titans, either confirming, rejecting, or modifying the *Why*. In this case, the audience is engaged, attentive, and the issue of affectivity is on the table right up front. There will be no wasted time now. As a presenter, pick up on any objections or discussions as to the *Why* and use those to modify your presentation to fit *Why* the audience thinks they are here.

Do not intentionally make the *Why* portion controversial. Like the wedding, one would expect that the *Why* portion is a given, is readily accepted by all. But, if it is not, it is best to know that up front so you can react, modify, and present accordingly. Or, as a minimum, do not waste their time, and thereby you live to fight another day. The *Why* segment of the presentation is usually not controversial, but if it is, then clearly the more reason to make sure all decision makers agree on *why* they are here. I have found from years of observation, however, that controversy is difficult to predict. It can attack unexpectedly.

I was scheduled to deliver a presentation on a highly classified project to about 150 government representatives and principals from other competing companies. It is unusual to have competing companies present competing technologies to the government-customer with everyone in the same room, but the government was in charge and that is the way it was. All the contractors had been working on this project, in separate compartments. for almost a year. The purpose of the entire three days of meetings was for the government to meld and integrate the set of technologies they chose from each contractor and make sure everyone was marching to the same drummer. Each contractor had been given guidelines.

It came my turn to introduce the technology my organization was working on. I stood, walked twenty feet to the eight-inch-high platform, faced the audience, introduced myself, and projected a viewgraph with the title of my briefing—the same title as on the printed agenda everyone held, and the same title as that given me by the government for this meeting. Instantly, the chief scientist from one of our competitors interrupted with an objection to the title and an objection to my organization giving the presentation (the latter being the real issue). He was a heavy man with his foghorn voice set to full volume. His comrades piled on in raucous agreement. Every contractor in the room took sides. (My enemy's enemy is my friend.) A food fight was on. If they thought the title was provocative, I could only imagine what was to come.

In spite of what appeared to be a catastrophe forming, I popped up the *Why* viewgraph. I was able to turn this strong confrontation into a great introduction of *Why* I was giving this presentation, *Why* all of these scientists and the government should listen, *Why* they should consider what I had to say, and *Why* they should enact my recommendations. Had I tried to start my presentation by telling them "*What I Plan to Say,*" my presentation would have added only another statistic to sudden crib death. As it was, the violent objections, and my countering with a *Why,* guaranteed that my customer, the government, would listen and that they would be ready to make a decision.

While it was not my original purpose, the use of the *Why* introduction brought all the controversy up front. The government settled the argument in my favor. The government decision makers, the only persons who counted in the meeting, were now, more than ever, eager to hear what I had to say. At the end of the three days, the government gave me a commendation for possessing a certain "grace under fire," a quote from Hemingway. My recommendations were accepted and that was the approach finally integrated into the finished system. But, the key point here is the effectiveness of using *Why*.

An additional example may help. Assume your presentation describes a new methodology for the evaluation of research and development proposals. You want the decision makers to no longer use the old methodology, but to approve and implement a new methodology you wish to propose. (You start with the *Why* they and you are here.) *Why* do they need to do this, they ask internally.

Why? "Because the existing process is indiscriminate and spreads the funds like peanut butter arbitrarily across competing business units. The existing process does not focus on the winners and cut the losers. This peanut butter approach to funding is a compromise in which no one wins. The projects that really need funding are stifled by a swarm of insignificant ones that overwhelm them when we apply this lowest common denominator financing. (Of course, different vice presidents disagreed on what were the most important projects, but that only added to the interest.) My analysis shows that the return on investment (ROI) could be improved by 10% if this recommended methodology is adopted." That's the *Why* we are here. To make a decision. Do you want to save money or not? Do you want to invest effectively? That will get their attention. Start with *Why*.

One of the best illustrations of the effectiveness of the *Why* approach came when I was working on another intelligence project. (I modify some portions, here, for security reasons.) Our team had developed a unique capability for correlating signals and tracking information and we were ready to show a demonstration to the government, the decision makers. We had been working with a government team, worker-bee intelligence wizards, for two months. Some top military brass, the bosses of our government team, flew in for the demonstration and for the simulation of the capabilities we had put together. The chief decision maker and his adjunct staff are seated in a large room, our system and computers are running in the background. The chief scientist walks to the platform, turns to the general and starts like this:

"General, the *Abu Simbel*, an Egyptian freighter, is scheduled to dock in Dubai within the hour. We discovered only last night that it made an unscheduled stop in Naples and loaded seventy tons of ammonium nitrate fertilizer. The waybill indicates it was destined for agricultural farms in Yemen." While the scientist was saying all this, related photographs were displayed on the large screen behind him, to which he feigned obliviousness. "There are three tractor-trailer trucks from separate warehouses currently bound for Dubai. We believe they will pick up the fertilizer. All are registered with the Jordanian company, *Fasil Hussain,* a known sponsor of terrorism. The trucks will arrive in Dubai long before the freighter is scheduled to depart on Thursday. We believe this is a terrorist shipment. We know all this because we, and your team, have developed a new intelligence capability. We are here, today, to determine if this capability is one you need, want, and one for which you will fund development." The scientist had delivered the *Why*.

Was the general interested? You bet he was, and the demonstration was a smashing success. The power of the *Why*. Admittedly, this dramatic approach was a little more dramatic than we usually are allowed, but the government team working with us felt it would be effective for this particular decision maker. It was.

Can you conceive the difference if we had started with, "Sir, we plan to tell you about some research we have been working on with your team that I think will be of interest to you. We have worked on this for the last six months. I will start by giving you a scenario of a simulated event. This scenario will also have pictures and other information that can be shown on our large screen behind me. The computer equipment that you see around the room is actually running the new software, written in C++, and processing through quad processors from Intel. After the scenario, I will show you schematics of the computer system to give you some idea of the processing capabilities required. Then, I will discuss the networks required to feed us the data. The heart of the matter, though, is the logic we use in assimilating the various telecommunication sources and the search algorithms to create a logical ontology for making decisions. That ontology is one we derived from earlier work we were doing with Boolean decision trees. Then, we will blah, blah, blah."

Now, at what point in this inferior approach, the standard one, do you think the general will check out and write us off? Correct. He will disconnect during the introduction, before he even understands what the meeting is about and why it is important to him. Decision makers must know *why* they are at the meeting, right away, and up front. The power of the *Why* when persuading decision makers is evident.

In using the *Why*, there are some errors that must be avoided. Do not try to make the *Why* portion "funny." Wizard comedy is seldom funny outside Wizard World. Likewise, do not make the *Why* portion trite, cynical, or clever. It is easy for a wizard to do this, almost natural, but it will neuter the *Why* segment to be superficial, reactionary, and counterproductive. Some examples I have heard, honestly, go as follows. Many of these were given in the most important of audiences. Here are the wizard openings.

"We're here to solve your problem and do it right, this time."

"I'm here because you asked me to be here."

"I'm not sure why I'm here, but my manager asked me to make this presentation."

"I'm here because I guess I drew the short straw."

"I'm here because my boss thought I didn't have anything better to do."

"My presentation will certainly improve the quality of this meeting from what I've heard so far."

Believe it or not, these are actual, bona fide openings. Unfortunately, I have heard many of them several times from different, unrelated technologists. Trust me, there is no underestimating the spontaneity of a wizard once some "clever" thought triggers a neuron.

The *Why* portion is not trivial. It is not lofty or rhetorical. It is a clear statement of the importance of each decision maker being at this meeting for this specified purpose. It is not a statement of why the speaker is here, but a statement of why the collective group, speaker and decision makers, has been brought together.

Start with a *Why*. Decision makers are not scientists, they are business and technology entrepreneurs. Focus like a laser. When you get on the same plane as the decision makers, you deliver to them the things they find important. Using the *Why* approach, you move them from the nonessential to the essential, from the description to the actual reason. Stick to the things that are of interest to the decision makers.

WHAT

The second step in the *Why, What, How* approach is the *What*. You present the essence of the work, the data, the charts, the graphs, the content, the arguments.

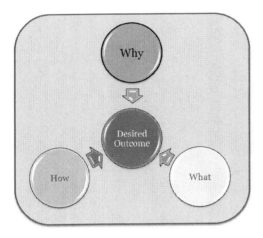

In the example of the marriage ceremony, the *What* portion describes the elements, duties, and responsibilities of marriage.

"Marriage is a holy estate and one should not enter into it lightly, for in marriage we join the two to the unity of one. Whereby the one may be weak, the two may be strong. Whereby the one may be strong, the two together may be stronger ..."

Plan the *What* segment to be take approximately 70% of the time you allocate for the total presentation. Note, the amount of time you internally plan for the presentation itself is much less than the amount of time actually scheduled for the meeting. We will discuss timing in detail in Chapter Seven, but one must allow for questions, interruptions, decision makers getting to their next meeting, and so on. For now, target the *What* segment to be 70% of the time you allot to speaking.

The *What* segment must contain sufficient, cogent evidence to elicit the action requested at the end. It must be comprehensive, compelling, and understandable by the decision makers.

As a general rule, most technology wizards create a *What* section of sufficient quality but err, exceedingly, in having far too many superfluous, technology details. Wizards cannot discard. In a desire to be comprehensive, they are exhaustive and frustrating, retaining detail after detail. For every point they wish to make, they fire a cannonball of techno-shrapnel instead of a rifle shot of decision-required information. The wizard *must* learn to differentiate between what is pertinent and what is peripheral—the crux of Chapter Eight.

The presenter of technology must look to information that speaks directly to her specific audience of decision makers. Maybe finance data is required, or information from human resources. If so, get some help from those departments. Do not try to imagine what financial charts are used by the company or what human resources information is reported. Just ask around and go to the department directly. Personnel from those departments usually appreciate your asking and usually are pleased to be of assistance. As simple as this sounds, most technologists, being introverted and loathe to ask for help, will waste an unbelievable amount of time creating something original that is inferior to the standard that already exists. Technologists would sooner use someone else's toothbrush than someone else's data. Engineers thrill to create and agonize to copy.

We expand considerably on the *What* component in subsequent chapters. Suffice it here as an introduction to the element.

HOW

In the *Why, What, How* approach, the wizard starts by telling the audience *Why* the meeting is occurring, why it is important, why everyone is needed here at the meeting.

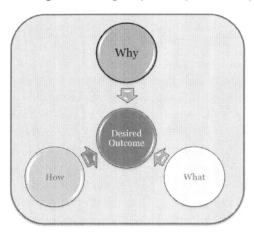

Then, you tell them the *What*, that is, what is the compelling evidence before us that cries out for a decision? After you successfully complete the *Why* and the *What*, you must close with the *How*.

Should you summarize, instead? No, no, no, no, no. You took valuable time from the most important people in the company and you want to spend your final minutes repeating what you said? No way. To repeat the *What* in a summary is repetitious, unnecessary, and decidedly ineffective. These are decision makers: highly intelligent, skillful, and keen listeners. They can hardly have forgotten what you told them just minutes ago. If you failed to explain it adequately in the *What* segment, you will be no more successful in a summary. They have a hardcopy right in front of them if they need a refresher. (The hardcopy is a handout you give them before you start. This is discussed in Chapter Eleven.) If they do not understand it at this point, you just did not make it clear and repeating it will not make it any clearer. Most presenters summarize because they need to end and do not know what else to do except to cycle through it all again.

Think of it this way. They all came into the room with their engines stone-cold dead. Your *Why* started the engines. You accelerated to top speed with the *What* segment. Now, they are almost ready to break the sound barrier and go boom, boom, "Yes, we agree with you; ask what you will." If you end your presentation with a summary, you might just as well slam on the brakes and smash everyone's head against the dash. Why would you do that?

Or think of it another way. You showed them the mountain with your *Why*. You climbed the hill with your *What*. You lead them to the precipice. You spread before them a universe of opportunities, the opportunities you articulated elegantly. And now, you want to crawl back to the bottom where you started, so you can take a nap? You want to summarize?

A summary is an echo of the real thing, hollow and empty. If the *What* presentation were a five-course meal, then the summary would be a belch at the end. Is that the effect you want? No.

What then?

It is time to call them to action. It is time for decision makers to—make decisions.

The final step is not to repeat the *What* and tell them *What* you just said. Rather, if you want to persuade, then end the presentation by telling the decision maker *How* to enact and enable the *What*. In other words, what to do now, how to implement, how to carry the technology ideas to fruition. How does one accomplish this? Show the impact that the *What* portion will have on their organizations, if they just would enact it. And in order to enact it, you tell them *How*.

Return to the wedding ceremony and see the *How* part in action. *"John, do you take Suzanne to be your lawfully wedded wife?" "I do"* (a decision to go forward with the *What*). *"Suzanne, do you take John to be your lawfully wedded husband?" "I do."* (a decision to enact the *What*). *"I now pronounce you husband and wife."* (a third decision to confirm). *"That which God hath joined together let not man put asunder"* (a formal seal on the decision).

If you have done the *What* portion correctly, the decision makers will anticipate the *How* portion. They will be ready to make a decision. The decision makers, John and Suzanne, have something to decide. They have listened to the *What* concerning marriage, what love is, what is important about the union, and what the commitment means. Now it is time to agree to the commitment. *How* do they implement the components of marriage? *How* do they take action? How do they advance from here? You must tell them. "Do you take this man? Do you take this woman? It is time for a decision."

This *How* step is critical in technical presentations to decision makers. You, the presenter, have convinced them (hopefully) that there is something worthy of their attention. Now you must tell them not only what they should do about it, but how they should do it. The *How* is essential, the *sine qua non*. Without it, why present to decision makers at all?

My team was preparing a presentation for the Secretary of Agriculture in a nearby state. We reviewed the presentation and were ready to print the final copies. We eyeballed our checklist. Yes, we had done a great job of telling the Secretary why we were meeting with her. Yes, we had explained in detail our products and what could enhance agriculture. Everything seemed perfect until we got to the end. In developing the presentation we had become so absorbed in communicating our greatness, that we forgot why we were there. Our ending was a resume of our qualifications. We asked ourselves, "What do we want with the Secretary? What decision did we want her to make?" We changed the ending to a *How* and described the endorsement and action we were requesting from the Secretary. It turned a ho-hum marketing overview into a decision-meeting.

Another time, we were working with the administrator at a nationally-known medical research hospital. The rest of my team flew up the morning before the presentation. I met with them in the hotel that night for dinner where we all reviewed the presentation that I planned to give for our team. Oops, we discovered that the presentation ended in a virtual summary. Here I was trying to sell an idea and we never asked for action. We were not salespersons, but we were technologists trying to sell a concept. We spent the next three hours deciding exactly what we did want from the hospital administrator. We surprisingly discovered that we did not know. We forgot *Why* we were there. We went back and examined our *Why* segment. We really did not have one of those, either. We were too busy talking about our accomplishments. We reworked the *Why,* and then were able to work the *How.*

Wizards are like that. We spend so much time looking at the technology that we sometimes forget why we get invited to these meetings. This structure of *Why, What, How* is not just an exercise in methodology. It transforms your thinking from looking internally to looking externally. It changes diffraction into reflection. It changes the entire intent and flux of your presentation.

I spent two winters working as an amateur archaeologist at Matagorda Bay in the Gulf of Mexico. We were searching for artifacts from the LaSalle expedition, near his fort that constituted the only French claim to Texas. Our search was through dense thorny brambles, six to eight feet high, and close enough to form a scarcely penetrable barrier. The thorns were the size and length of toothpicks, about two inches. The task was to comb our way through the underbrush on our hands and knees looking for probable artifacts. The thorns tore at your jacket, grabbed at your headgear, and pierced through your hands and knees.

To do the search, we started out in the field, side by side, about ten feet apart, and then began the crawl through the brambles, headed towards Garcitas Creek. Inside the maze of interlocking brambles, it was easy to lose the way and get off track. It was with a real sense of victory that you finally reached the river and, like Brer Rabbit, could boast that you had traversed your part of the briar patch. The problem was, that in trying to get to the river, you would forget why you were there. You would forget that you were there to look for *artifacts*. That, after all, was the *Why*. The *Why* was not to get through the briar patch to the river, but to find artifacts.

You see, it was easy to forget *Why* we were there. We had to keep reminding ourselves so we would remember that the goal was not to get to the river, but to make discoveries. The same way in a presentation. Make sure you remember *Why* you are there. You cannot be successful in telling a decision maker *How* to implement, if you forget the entire purpose, the *Why*. Make sure that the *How* closes with the *Why*.

Spend 10% of the time telling the decision makers *Why* in the presentation. Spend 70% of the time making a case and describing the situation. Spend 20% of the time with the *How*.

In practice, articulating *How* is difficult. Most technology presentations end effetely. I hear hundreds of presentations in which the *What* portion is compelling, but the presenter closes with, "Well, that's the end of my presentation. What would you like to do?" Or another time, "I would like for you to implement these ideas (no clue as to exactly what or exactly how)." Well, of course you would. Why else would you be here? That does not tell the audience *How* to implement.

Even something as straightforward as a status report to management can have a powerful *How* ending, informing what the next steps are, and *How* the audience can implement them.

The *How* segment cannot be a statement of the speaker's wishes and desires. It cannot be a "To-Do" list of how to implement, and of "what next." It must be a compelling argument so that the decision makers themselves see the cause and *decide* in your favor.

Now, there you have it. The *Why, What,* and *How*. But, you ask, is it really correct? Is it really the right structure to use with decision makers? Is it truly that good? Yes.

After I painstakingly discovered the success of the *Why, What, How* approach, I researched the speeches of world-renowned orators. I wanted to see what approach they used when they needed to convince decision makers to make a decision. I was stunned that they had beat me to it and already knew the overwhelming effectiveness of the *Why, What, How* structure. Here are a few vignettes from that discovery.

Shakespeare's *Henry V* gives a summons to battle, thought by critics to be historically accurate. Whether the speech is rhetorically exact or not, the essence is there and the greatest English writer of all times showed the power of the *Why, What, How*. Here is the background.

The army of Henry V is outnumbered five to one at Agincourt. The King makes an appeal to his ragtag, exhausted, and discouraged infantry. He uses the *Why, What, How* approach and ends by telling each man (decision maker) *How* and in what direction their decision is to be made. Can he convince them with the *Why, What, How* approach? Listen and see. His cousin, Westmorland, introduces Henry's speech.

Westmorland: "O that we now had here but one ten thousand of those men in England that do no work to-day!"

[King Henry starts his *Why*]

King Henry V: "What's he that wishes so? My cousin Westmoreland? No, my fair cousin; if we are marked to die, we are enough to do our country loss; and if to live, the fewer men, the greater share of honor. God's will! I pray thee, wish not one man more."

(This is the *Why* portion. We are here for honor, we, all of us. That is the *Why* we are here.)

[Now Henry moves into his *What*]

"By Jove, I am not covetous for gold, nor care I who doth feed upon my cost; it yearns me not if men my garments wear; such outward things dwell not in my desires. But if it be a sin to covet honor, I am the most offending soul alive. No, faith, my coz, wish not a man from England. God's peace! I would not lose so great an honor as one man more methinks would share from me for the best hope I have. O, do not wish one more!

Rather proclaim it, Westmoreland, through my host, that he which hath no stomach to this fight, let him depart; his passport shall be made, and crowns for convoy put into his purse; we would not die in that man's company that fears his fellowship to die with us.

This day is called the feast of Crispian. He that outlives this day, and comes safe home, will stand a tip-toe when this day is named, and rouse him at the name of Crispian. He that shall live this day, and see old age, will yearly on the vigil feast his neighbors, and say 'To-morrow is Saint Crispian.' Then will he strip his sleeve and show his scars, and say 'These wounds I had on Crispian's day.' Old men forget; yet all shall be forgot, but he'll remember, with advantages, what feats he did that day. Then shall our names, familiar in his mouth as household words - Harry the King, Bedford and Exeter, Warwick and Talbot, Salisbury and Gloucester - Be in their flowing cups freshly remembered. This story shall the good man teach his son; and Crispin Crispian shall ne'er go by, from this day to the ending of the world, but we in it shall be remembered - We few, we happy few, we band of brothers;

[At this point, Henry tells them *How*—*be willing to shed your blood, today and fight*]

For he to-day that sheds his blood with me shall be my brother; be he ne'er so vile, this day shall gentle his condition; and gentlemen in England now-a-bed shall think themselves accurs'd they were not here, and hold their manhoods cheap whiles any speaks that fought with us upon Saint Crispin's day."

This is a powerful presentation—*Why* we are here (honor), *What* we can do about it (fight), *and How* we should do it (fight today for King Henry). Note that Henry moves seamlessly through his *Why and What* to his *How*. No break in rhythm, just here is *Why*, here is *What*, and here is *How* you can enact it. This *How* step is essential in a presentation to decision makers so that they understand exactly *How* you want them to proceed. Do you want them to fund a project? Do you want endorsement for a new endeavor? Do you want staffing? If you have done your work well, they are eagerly anticipating this *How* step because they live for decision, action, and implementation.

Go back and think about Henry's speech. See how weak it would be with the *Three Whats* approach. True, he is not talking about technology, but his destiny hangs on the decision of his soldiers. He wants every soldier to love honor more than life. That choice is the decision he forces them to make with the *Why, What, How* approach. Powerful.

In 1940, at a most crucial point in the war, Winston Churchill took office as Prime Minister of Great Britain. In his very first address to Parliament his absolute first sentence starts with *Why*.

[Why are we all here?]

"On Friday evening last I received from His Majesty the mission to form a new administration."

He is telling them *Why* they are all here and *Why* the meeting is important to all of them. He goes on to tell them *What* he has done, appointed a war cabinet, appointment of other ministers and other such specific things. Does he end here with the *What*, or by summarizing the *What*? No, not by any means. In fact, a full one-third of his presentation is taken up with the *How* because he needs to lay out to the decision makers of parliament exactly *How* they are to implement the *What*.

[Churchill's *How*]

"You ask, what is our policy? I say it is to wage war by land, sea and air. War with all our might and with all the strength God has given us... You ask, what is our aim? I can answer in one word. It is victory. Victory at all costs... Come then, let us go forward together with our united strength."

One year before the US declared its independence from Britain, Patrick Henry stood before the Virginia Convention of Delegates. What structure did he use for his impassioned speech to get the delegates to join the cause of freedom? You guessed it: *Why, What, How*.

[Patrick Henry's Why am I giving this presentation and Why you should listen]

"Mr. President, ...I shall speak forth my sentiments freely and without reserve. This is not time for ceremony. The question before the House is one of awful moment to this country."

Mr. Henry continues telling them about the *What*, that is, what has occurred in the colonies. He wants a decision from the decision makers. He knows the power of the *Why, What, How*. Here is his finale, the *How*.

[Patrick Henry's *How*]

"...there is no peace. The war is actually begun. The next gale that sweeps from the north will bring to our ears the clash of resounding arms! Our brethren are already in the field! Why stand we here idle? ... Is life so dear, or peace so sweet, as to be purchased at the price of chains and slavery? Forbid it, Almighty God! I know what course others may take; but as for me, give me liberty or give me death!

(Meaning your action must be the same. Patrick Henry casts his vote as a decision maker and then challenges all the other decision makers to follow him.)

Note how effective the *How* portion becomes when it is not simple rote or desire, when it is woven into the fabric of the *What*. Few wizards can reach the pinnacle of these impassioned speakers, but we can learn a great deal from observing how they successfully persuaded otherwise recalcitrant decision makers.

But let us move directly to wizards facing decision makers, our target. No one can argue that possibly the greatest wizard of the last two centuries was Albert Einstein. After the bombing of Hiroshima, Einstein advocated world government. His speech was broadcast across the nation. I comment not on his proposed plan, but let us examine what method he used. The underlines are mine.

[*Why* I am making this speech and *Why* all you attendees should listen]

"*I am grateful to you for <u>the opportunity to express my conviction</u> in this most important political question. The idea of achieving security through national armament is, at the present state of military technique, a disastrous illusion.*"

He then elaborates on the *What* portion which does not concern us at the moment. In the *What*, he discusses the issues. He closes his speech, of course, with the *How*.

[Here is *How* you, the citizens and decision makers of America, can implement this plan I propose]

"*Is there any way out of this impasse created by man himself?... The first problem is to <u>do away with mutual fear and distrust. Solemn renunciation of violence is undoubtedly necessary</u>...*"

Einstein's use of the *Why, What, How* is not as effective as some of the other speeches, at least to me, but English was a second language to this brilliant man. Also, consider that he was extremely shy and not a gifted speaker in any language. In short, he was, well, a *wizard*. But he knew how to persuade decision makers.

Technology wizards think their job is finished when the facts and data are presented. They think their responsibility is complete and the next step should be obvious to even the casual listener. It is not. This is a mistake of magnitude. One can spend hours articulating facts, figures, and data, but you will never get action until you tell them the action you want them to take, and *How* to do it. You will be surprised at the number of technologists who do not know what they actually want from the decision makers. They just get to the end and peter out.

Decision makers are busy. They have not the time or the inclination to connect all your dots. They are called decision makers because they make decisions and this is what makes the structure of *Why, What, How* so effective. It plays to their strengths. It plays to your strengths. You combine your strengths with their strengths. But you cannot do this successfully unless you show them *How*, exactly, to couple their decision-making capability with your technical acumen. You are the wizard. Do it.

This three-step process, this *Why, What, How* is the optimum approach for decision makers. Tell them *Why* they are here. Tell them *What* they should consider and need to know—the facts, the data. Tell them *How* they enact their decision, how to implement it, how to ensure its outcome.

Right now, most wizards are probably saying, "I have the concept, but that does not tell me the implementation." Knowing the framework is just the beginning, as you shall see.

Paradigm For Success

o The *Why, What, How* paradigm is the most successful for influencing decision makers.

o Proven success over the centuries for all types of decision makers in all types of environments.

o *Why*—why I am here, why you are here, why we are having this meeting.

o *What*—what it is I have to say that is important to you.

o *How*—how the audience can enact and implement your recommendations.

Chapter 6: Strategic Plans
Designing a Favorable Outcome

*Opportunity is missed by most people because it is dressed
in overalls and looks like work.*
— Thomas Edison

A persuasive technical presentation is a significant accomplishment. The hurdle is high. You must posit so compelling an argument and so logical a recommendation that "do it" is the only decision possible. Your next presentation can accomplish this. But, to get to that point, we have much work to do. There is no easy path. No one creates a persuasive argument effortlessly. The ideas may originate in epiphanies, but to organize them into a persuasive force takes unbelievable amounts of time and energy. No great accomplishment comes without effort. This one is no different.

The delivery of the presentation takes only a few minutes, but the effort to do that successfully can be enormous. That final event in the conference room is only the tip of the iceberg. A persuasive argument is built upon a foundation of education, experience, analysis, formulation, knowledge, and conclusions. These technical presentations to decision makers are not for the faint-hearted or torpid. For every hour you spend making the presentation you need to spend at least ten hours in preparation. And for every level that the decision maker is above you, you should probably multiply that ten hours by the number of levels. In other words, if the president is four levels above you in management, plan on spending about 40 total personnel hours preparing the arguments and the presentation.

How does one create a persuasive technology presentation?

The following steps will result in successful persuasion. These steps must be taken and each one completed successfully before you stand in front of the decision makers. Items 1 – 7 will be covered in this chapter. Items 8 – 14 are covered in subsequent chapters.

Strategic Steps

1. *Start at the finish. Precisely define the end result you wish to accomplish. What decision do you want from the decision makers?*

2. *Work backwards to the data. Be certain that the data logically and uniquely point to the end result.*

3. *Develop a strategy. (The data obviously lead to the conclusions, but what thread of logic best explains that to a non-technical decision maker? We call this the strategy.)*

4. *Prioritize the key elements of the strategy.*

5. *Make a complete draft of the presentation. (A complete draft must include text, graphics, figures, pictorials, and drawings.)*

6. *Shore up any deficiencies in the data, analyses, or conclusions.*

7. *Tailor your presentation and strategy to the specific audience.*

8. *Pare and compress the draft to fit the amount of time you will present.*

9. *Perform peer reviews. Make modifications to the data, strategy, logic, key conclusions, and the graphic illustrations. Avoid text modifications.*

10. *Dry-run the presentation with a knowledgeable team, the so-called Pink or Red Teams. Look for holes in the logic first. Make text changes last. Convert all negative statements to positive statements.*

11. *Edit the text and finish. Perform spelling check, correct grammar (number, tense, and similar), ensure format consistency, and adhere to policy convention and copyright permissions where applicable. Make final copies.*

12. *Memorize and practice your opening and closing lines.*

13. *Reconnoiter the lay of the land.*

14. *Secure the underpinnings. Brief the staffers and subordinates of the decision makers. Brief any other technical experts who either factor into a favorable decision or represent strong controversy.*

START AT THE FINISH

Steven Covey popularized the notion of starting with the end in mind. It is the right concept for our situation. Just as you would do in planning your travel, you first must

> *Start at the finish. Precisely define the end result you wish to accomplish. What decision do you want from the decision makers?*

decide where you want to be before you can determine how to get there. It is also a great help to know where you are when you start. The technical presentation is not so straightforward as a travel plan, but it is necessary to know what you want out of the presentation before you can develop a strategy for getting to it and before you have any chance of actually getting there.

Your presentation answers some need. Otherwise, why give it? What is that need? If your presentation is the answer, what is the question? Think of it like the game show, *Jeopardy*. You have the answer—your presentation. But, what question does your presentation answer? What question are the decision makers asking, for which you have the answer?

What is it, exactly, that you want the decision makers to decide? What would constitute a successful meeting? (The response, "to come out alive," has been used already. Try another one.) What one thing do you most want the presentation to accomplish? This will be the *Why* element of the *Why, What, How* paradigm. Decide, first of all, what question you are trying to answer. *Why* are you there, really?

Decide the end result. Decide what you wish to accomplish. Do this before you get any data or look at any analyses or formulations. What will be the end result?

What confirmation, what commitment, what endorsement do you want from the decision makers? Do you need to come away with R&D funding for a project? Do you need approval for creating a new product? Do you want to change a process in the production line? Do you want to run more tests on the new drug? Do you want to get government approval? File for a patent? Get a raise? Get a promotion? Sell an invention? Form a new organization? Modify the organization? This "desired end state" requires more time to formulate than you might think. The end result is initially difficult to articulate because you need it to be succinct and to quantify. Once you exactly state the end result you take one giant step towards persuasion. You know what you want.

If your end result sounds something like, "I want to report status on my recent work," or "I want to show the latest test results," or "I want to make sure they understand what I'm working on," then, stop. That is not an end result. You are not making a presentation, you are copying a Day Timer calendar. There is no reason to meet with a superior just to show them you have been working and doing what they expected. Perhaps you respond, "But I have been asked to give status of the project. Is not that my end result?" No. No level of decision maker is interested in how busy you have been or how many hours you have worked. Their job is to *make decisions*. They do not do experiments, they do not do research, they *make decisions*. Therefore, you must give them something to cogitate and decide.

It is assumed already that any professional is working sufficient hours and at a sufficient pace. If that is in question, it is a different meeting and not the subject of this book. Decision makers are not concerned about whether you are working diligently or how many hours you are spending (except as measured by results and output). They are concerned about whether you are on schedule, whether you have a schedule, whether your schedule matches their schedule. They want to know if you are staffed adequately. Do you have the right talent or caliber of staffing? Do you know what your budget is? Are you on budget? Do you even have a budget? When will you be finished? What is your definition of being "finished?" What will be the artifact, or output, that will prove you are finished? Is it a report? Is it a patent? A new policy? A different procedure? What are the next steps? What are the decision points? Do you have key milestones? Have you reached them? These are the types of questions that decision makers mean when they say "Give me your status." And the end result is not, "I am on budget. I am on schedule. I have adequate staffing." What do you want them to decide? To continue funding? To increase funding? To add staffing? To change the production forecasts? If there is not something for them to decide, judge, or evaluate, why are they meeting with you? They are not potted plants.

It may help to illustrate. Look into a courtroom and view the prosecutor at work. Does she stand before the jury to describe how hard her team is working, how she is on schedule, or how she manages her budget? No. She makes a case. She brings forth the evidence, the facts, and the circumstances, and then influences the judge and jury (read: decision makers) to make a decision. Her end result is to get a conviction. It is little succor to the prosecution if the jury commends her abilities and admires her acumen, but does not give her the verdict she requests. The verdict is the end game, the end result. The prosecution is after the correct verdict. As a technology presenter, like the prosecutor, you must know what it is you are trying to accomplish. You must know the question you are answering.

So what is the end result you want? Perhaps the results of your experiments are finalized and you believe your new product is ready to go to market. Maybe the test trials for the new drug are complete. Maybe you want approval and funding to file a patent application. Maybe you had an idea for a new business and put a draft plan together, but you need authorization to proceed. Maybe you want to convince an entrepreneur to invest in your idea. Maybe you found critical bugs in the new software release and need to work damage control with customers. Maybe you want a promotion or a salary increase. All these scenarios require a compelling, decision making, technical presentation. Know your end result.

What specific action do you want the decision makers to take? Distill it down to a single sentence or a single phrase. Decant the adjectives and adverbs and make it subject, verb and object. Next, add the quantification, the specifics of how much and when. The more exact the quantification, the better. An example is, "I want an entrepreneur to invest $250,000 to complete this research within the year." Or, "I want the vice-president to agree to market this product by July 1." Or, "I want my manager to add three medical technicians to my department so that my department can do 'A,' 'B,' and 'C.'"

This is not as easy as it may appear. As you begin this process, and I have done it for years, you think you know why you are meeting and what you want out of the meeting. But, once you put it into words, do not be surprised if you say, "No, that's not right. That's not exactly what I want out of this meeting." It usually takes three to five iterations to get it right.

WORK BACKWARDS TO THE DATA

You know where you want to be. Do you know where you are? You cannot answer, "Nowhere." Quantify and specify.

> Work backwards to the data. Be certain that the data logically and uniquely point to the end result.

In 1878, Edison knew what he wanted. He wanted to invent a commercially practical incandescent light. He knew his end goal exactly. But where was he? His team had been working on this unsuccessfully for years. He could have said, "I'm nowhere." He could have said that, but he did not. He knew where he was. He said, as variously quoted, "I have not failed 700 times. I have not failed once. I have succeeded in proving that those 700 ways will not work. When I have eliminated the ways that will not work, I will find the way that will work."

Quantify where you are in your research. Quantify where you are in trying to get funding for a project. Quantify where you are in your quest for advancement. Whatever the subject, quantify it. If you are seeking promotion, quantify your education, your work history, your successes, your failures, your assets, and your liabilities. Be candid. You have to know where you are before you can plan how to get somewhere else.

You want a presentation that gives the end result, B. You know where you are, at A. What data do you have that bridges those two? This is the only data that matters. Look at that data only. Work backwards, then forwards. Do the data lead to the conclusion you want from the decision makers, or are there many conclusions that could be drawn? To the extent possible, you want a one-to-one mapping from where you are to where you want to be. You desire only one logical conclusion from the data. You want the decision makers to see it and make that one decision.

Do the data logically and forcefully lead from where you are to where you want to be? Check again and make sure. Examine the data and their analyses. This is the evidence upon which you will base your case to the decision makers. Are the data sufficient? Are they the correct data to answer the question? If not, make a list of what is missing and go to work getting the additional supporting information. What conclusions would an independent consultant draw from this same data? Would she confirm your conclusions? How strong is your case? It is not uncommon to discover you do not have a case, or at least not a solid one. Examine your conclusions carefully, because this is exactly what the decision makers will do. Do not fall into the wizard trap of thinking that more data means better conclusions. The question, the data, and the answer must align. The data and formulations from the data must compel the decision makers to a single conclusion. The decision makers will approve or reject based on this data.

In the criminal law analogy, the district attorney hears the case from law enforcement and determines if there is sufficient evidence to make the case, to go forward with a grand jury, and to make arrests. The district attorney looks at the evidence and says, "Can I win this one in court?" If the decision is "No," then he will quantify and specify what additional evidence is needed to make the case. You must do the same. If there are holes in your argument, you must get data to fill them.

If you cannot get data, or if the data do not lead to the conclusion you want, then reformulate the desired end results to match the data. Alternatively, go get new data that will substantiate the end results you want. Make absolutely certain that you can go from point A to point B with the data you have. Do this in the very beginning of your strategy, not two hours before the presentation. Do the data, themselves, lead to a unique, indisputable conclusion? If not, then modify your conclusions and your recommendations. Modify B. Do not dishonestly change the data to make them say what they do not. Recall our discussion, earlier, about trust? Trust starts with being honest with yourself. The data are what they are. Do not pick and choose data to get the end results you want. Do not try to put a "spin" on it. The point is to make sure all the data give the right answer to the right question. If they do not, then you should reformulate the meeting, change the conclusions, or change your recommendations. I have done this many times. Be honest.

DEVELOP A STRATEGY

At this point, you have determined that the data logically lead to the conclusion. For example, you need two new researchers in order to complete your project within six months. Your data show the quantity of work to be done, how much each

Develop a strategy. (The data obviously lead to the conclusions, but what thread of logic best explains that to a non-technical decision maker? We call this the strategy.)

person can be expected to do, and how many people that takes. The data show the qualifications of each person. Excellent. The logic is sound. The data lead from point A to point B. But what strategy do you use to communicate all that? You cannot just proclaim, "Here's the data, and these are my recommendations. That's it. Take it or leave it." I assure you, they will leave it.

The decision makers are going to ask obvious questions like, "How do you know how much research is still required? How is it you can predict research?" These are common questions and difficult to answer.

Most wizards would say, "I can't. There's no way I can predict research."

But that leads to the obvious counter by the decision maker, "Then, how do you know you need two people? How do you know you need any? How do you know you don't need six instead of two?" Or, worse, they may say, "Then, why don't we get someone in here who does know?" So, there you go down a path that will lead to nothing good.

Thus, you must develop a strategy to travel from point A to point B. You have to explain it all to non-technical decision makers. You must learn to think like a decision maker and ask these questions to yourself ahead of time.

Here is another example of why you need to develop a strategy. Suppose you are meeting with entrepreneurs and requesting they invest six million dollars in your new idea. You, the technologist, think it is a great idea, probably the next Microsoft, Google, or Amazon. A great idea is not enough. Maybe you have gone to the next step and put together a *pro forma* financial statement. (A *pro forma* is a spreadsheet of time-based, theoretical, proposed, financial transactions for your product.) What strategy will you use to convince them to lend you six million dollars? Will you just tell them, "Trust me?" Probably not. Will you start with the *pro forma*? Maybe. Will you start with a description of the product and its impact on the market? Maybe. The "maybes" are why you need to develop a strategy. You cannot just go into the meeting and wander around.

I have a solid technique for developing a strategy. It is just a technique, but it works. I recommend you, alone or with your team, get in front of a large white board. I know I offend every techie, but I do recommend specifically that you not work the initial strategy on a computer. Hear me out. First of all, I have used a computer my entire career. I can type 60 words-per-minute. I compose almost everything I do in real time on the computer—the draft of this book, documents, presentations, calendars, calculations, spreadsheets, schedules, and programs. I multitask between text messages, emails, videoconferencing, Word, Excel, PowerPoint and the like. Yet I find existing software to be far, far too *slow* to use to develop strategy, and I have considerable computer and office skills (broader than most scientists or engineers). Computers are not yet to Bill Gates' "business at the speed of thought." Thinking is free-hand, requires bouncing around, quick iterations, up and down, walking about, a try this, try that.

Computer wizards might argue that a laptop, videoconferencing, or a portable projection permit a small group to *create* ideas faster and synergistically better than any other technique—all can see, all can contribute. I argue it is simply a force-fit, a square peg in a round hole. One person at a time controls the computer and, if you are a team member not controlling the computer, it takes too long to either seize control or explain your idea to the team member operating the computer. Everyone tries to edit as you go. Graphics are interminably slow and inadequate. Sketches are cumbersome, at best. It takes far too long to even hope to keep up with the brain. It inhibits follow-on ideas. It quickly degenerates into "wordsmithing," making trivial changes in wording before you even know what you want to say. If the team is participating from remote physical locations, real interactive *creation* by the team advances at about the speed of the post office queue. The commercial software for interactive meetings, with participants at remote locations, is functional—I use those programs every day. But they are marginal tools for *creating* technology strategy. The operative word is *creating*.

I recommend you *create* the strategy and the connecting logic, whether alone or with a small group, using the white board. You can pace around, erase this, add that, change colors, sketch, draw, three people dueling with markers at the same time, erasing and writing at the same time, scratching through things. It works at the speed of thought. This process works faster, and in more "dimensions" (various capabilities) than is currently allowed by the computer. Until we get several generations down the road in interactive software, I strongly recommend doing it the old-fashioned, but better and faster way. Optimize the use of tools.

I possess a trade secret disclosure for equipment to measure ink-jet printing head quality. When we designed the prototype equipment and built the sensors, we had the idea of using multiple lasers to measure dynamic ink-drop shape from the print heads. I purchased the latest and greatest in small and expensive lasers. The whole setup *looked* really techie. But it failed to work as I designed and theorized it. We discovered that those very expensive lasers were creating near-field wave interference problems that were reducing the accuracy of our measurements. We replaced the high-tech lasers with ten-dollar diodes and everything worked like a champ. But it took several weeks for me to make this decision, not because I could not identify the problem, but because I could not force myself to depart from an advanced technology. It is difficult for a wizard not to use the latest gadget, even when suboptimum. I stand by the white board recommendation for developing strategy. Whatever mechanism one uses, though, getting started is the hardest part. It is the place where most technologists experience something discussed earlier, *aporeo*—totally at a loss for what to do.

How does one get started on a strategy?

Outlining. Some texts recommend outlining. I do not. An outline puts key elements in order. When you start, you do not know the elements or the order. Now, you might argue that you could use word processing software, putting down thoughts, shifting them around, automatically numbering, and so forth. You could argue that, even if the thoughts are random, they can be organized and put in order. In spite of that argument, outlining is an effete tool for *creation*. Period. Once the creative juices start to flow, the outline slows thoughts and logic to a standstill. The outline is too structured for creative thought. You change this, you change that, one thing affects another and you stalemate in short order. The problem is that the outline approach is a linear sequence of thoughts and strategy is multidimensional. The outline, if you produce one at all, is the last thing you produce, not the first. You create the multidimensional strategy, first, and then project it into the one-dimensional space of an outline. You produce the outline *after* you know the strategy you want to use, not before. Personally, I never produce an actual outline at any step.

Brainstorming. Maybe that's it. We all sit around the room and come up with ideas. Brainstorming is inadequate, too. First, it does not put the thoughts together logically. Second, the team bogs down in interactive critique. Third, if a "no critiquing" rule is implemented, you wind up with a disconnected, redundant collection of nothing.

I attended a workshop where brainstorming techniques were taught. My fellow teammates were mostly medical personnel. The facilitator posed the problem, "Devise solutions for preventing or mitigating pattern baldness." "No critiquing," he added. We all started yelling ideas while the facilitator wrote furiously on a white board, "toupees," "wear a hat," "implants," "transplants," "scalp replacement," "shoe polish," "Preparation H," ...

The facilitator stopped and help up his hand. "I know I implemented a 'no critiquing' rule, but I have to ask, 'Is there some chemical in shoe polish that helps prevent baldness?'"

The respondent said, "No. The shoe polish is to paint your head black so you won't look so bald."

"And Preparation H? How does that help?" the facilitator queried.

Another creative thinker spoke up, "You rub Preparation H all over the top of your head. The skin shrinks together and pulls the hair on the sides up to the top."

Such is the power of brainstorming.

Spider (Random Action Convergence). I recommend a multidimensional approach whose end product is a *Spider Diagram*. The *Spider Diagram* may have other names. It builds its strength on the convergence of random actions. It gets called a spider because the final result might possibly be imagined to look like a spider's web. (If so, it is either great imagination or a psychotic spider.) The *Spider* has been around a long time, in various incarnations. Programs like Microsoft Word have the *Spider* as one of their diagrams. However, as I stated earlier, these are acceptable for printing out a *Spider* but are diabolically slow and useless for any creation of ideas. Here is how to use the *Spider*, with effect.

Draw a circle—it does not have to be a pretty circle, but a circle, nonetheless, and only a circle. Do not draw a rectangle, polygon or ellipse. Draw a circle, right in the center of the board. Why a circle? Because rectangles, polygons, and ovals have sides and orientation, there is an up and down, a left and right, and sides imply a limit and direction. The circle has none of these inherent and restrictive aspects. Why are these other objects restricting? Because, as you put ideas and logic in place, you do not want to be restricted to a preferred order or hierarchy. Creation is a multidimensional space. You need a way to capture that space, the *Spider*.

In the center of this circle put your desired end result. This is the decision you want to come away with. This is the *Why* you are having the meeting in the first place. Now you have it—a circle with the end result written in the middle.

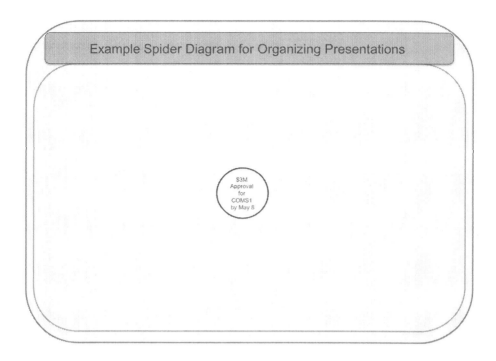

Figure 6-1: The Beginning Spider

As an example, let us assume that your research has discovered a mechanism for a new communications device. In our example, we will call the planned device, COMS1 (communications device number one, let's say). Let us assume the purpose of the meeting is to get additional funding to validate COMS1. Now, the hard part. How much and by when? This goes back to step one, *Start at the Finish*. *When* do you need the money, and *how much* money do you need? Make this precise. If you cannot do that, then you are not ready for the meeting. In the central circle you write, *Get approval for $3M additional COMS1 research funding by May 1.* (Write small, think big.) Refer to Figure 6-1.

What next? Think of the data and results you have, *any* evidence, *any* logic that comes to your mind, *anything* you think of that you will need to make or substantiate your argument. Call this the first thought. Continuing our example, let us assume that your first thought is this: *the results of the initial tests show that an increase in data transmission speed is possible with the new design.* Draw a line from the main circle in any direction. Put another circle at the end of this line and write in this new circle the data, evidence, or logic constituting your first thought. In our example, we write *data imply transmission speed improvement.* How much is this improvement? Well, you estimate about a 50% improvement. Erase what you have in the second circle and put in the 50% number. Now, your spider diagram looks like this.

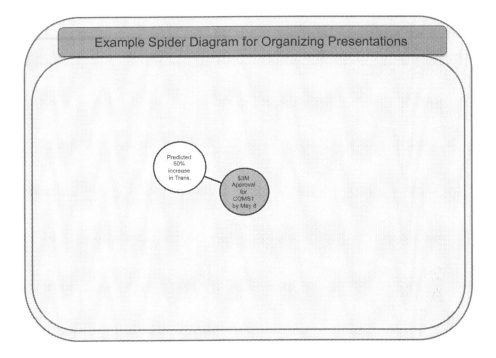

Figure 6-2: The Spider Grows

Quickly, what else comes to your mind, or to the mind of your team? This will be your second thought. If this second thought is completely unrelated to the first thought, draw another line from the initial circle and repeat the above. If the second thought is related to the first thought, then draw the line in any direction from the circle containing the first thought. In our example, let us suppose the next thought is that additional staffing would be needed. This is more related to the end result than the improvement, so, make a circle anywhere around the end-results circle and write in the circle, *Add one COMS expert*.

Then someone says, "We also need to add two additional programmers." Since this is still about personnel, you reason, this third thought *is* connected to the second thought, so you draw a line from the second thought and add another circle saying, *Add two C-Programmers*. Do not fret and argue about connectivity at this point. Just do it quickly. The creative juices are flowing. Do not hinder them. Our *Spider Diagram* now looks like Figure 6-3.

This process gets repeated over and over. If thought number eight was initially connected to thought number seven, but is also related to thought number three, then draw lines connecting the circle of thought eight with the circle of thoughts seven and three. If one connection is stronger than the other, maybe you could make one dotted and the other solid. Maybe different colors to signify different departments approving

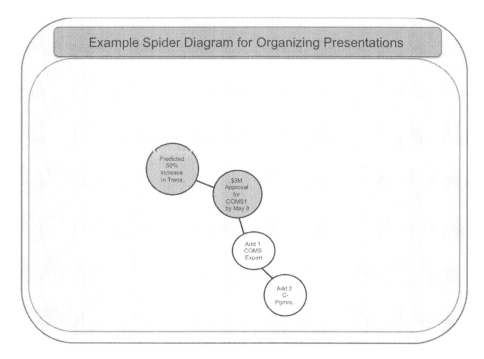

Figure 6-3: Spider 3

the step. Whatever. If these relationship have more complex distinctions, then put a label on the line that connects the two thoughts.

In our example, let us add a few more thoughts and see how it looks. One of the team members thinks that a new prototype must be built to test the result. Add that. Another team member estimates about $1.5M of the $3M request is capital equipment alone. The rest is labor. Another team member estimates that four months will be required to get the new data. You add that while you do the research, you may as well apply for a new FCC license since that process takes months. Now, what is driving the May 1 date? It is this presentation. And, the date of this presentation is driven by the board meeting date. But, in order for the board to approve the funding, the business plan must be developed two weeks prior and approved by finance. You add all these things and continue the process. Maybe it looks something like Spider 4.

What this *Spider* process does is that it captures the thoughts without inhibiting the creative juices. If you make a mistake, or change your thoughts, erase and move the circles around. Add dotted or different colored lines, if that helps. Regroup items. Erase some, add others. Within minutes, the entire presentation *logic* is mapped out. The thoughts were at random, scattered, disjointed, but now they are connected. Now, there is a pattern.

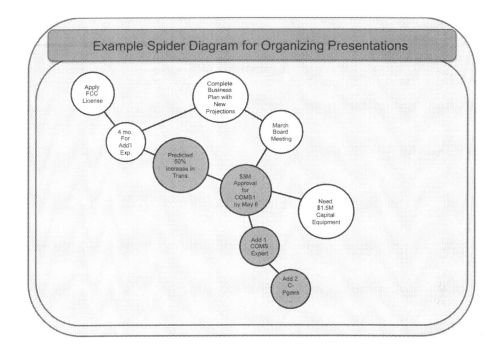

Figure 6-4: Spider 4

You may be working with a team or solo. It does not matter. If it is a team effort, everyone usually has access to all the markers. You have to fight and argue to get a point on the board, but do it quickly. If there is contention, just put up the thought anyway. You can argue it later. If it is total nonsense, have the person put it to one side of the *Spider* and bring it up later. Above all, keep it going as fast as your brain.

Now, you come to a lull and no more good ideas spring forth. Look at what you have. You should clearly see any holes in your presentation. There are usually several of those. Let us say you made estimates of the numbers. Then, those must be replaced with calculated values. Perhaps you now see that a business plan is needed. Or, maybe you identify that you have a weak business plan. Perhaps you should assign a team, or some individual, to create a realistic business plan. You need it well before the meeting because it may impact everything. Set a date for when you need it. Someone points out that the request for capital must fit the company's capital budget cycle. Is that a problem? Maybe you need to add some more steps in your *Spider Diagram*.

Note that the *Spider* process is fundamentally different from brainstorming. With the *Spider*, the thoughts are random, but they are directed at the end-result and linked to other related ideas. Brainstorming ideas are not initially critiqued. *Spider* ideas are dynamically linked, and given labels with some measure of importance—the farther they are from the central circle, the less important they may be. The more connections you have to the end-result circle, the more points you will need to emphasize in the presentation. Refer to Figure 6-5.

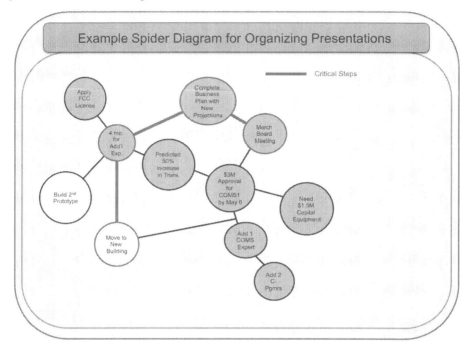

Figure 6-5: The Completed Spider

The point is, that just by approaching the presentation in this manner, the technology itself is improved. All those other things you once thought incidental to the real work are not incidental at all. You now see how it all plays together. You are thinking like a decision maker thinks.

Determine if there are any holes in either the data or the conclusions. Mark those holes or voids as things that are absolutely required before you can stand in front of the decision makers.

The *Spider* Diagram is your first cut at a strategy.

PRIORITIZE THE KEY ELEMENTS

What is the next step? You need to refine ideas, correlate them (check the linking), and eliminate redundancy.

Prioritize the key elements of the strategy.

First, eliminate the redundancies. There will be items that are actually repeats on some level. They are not worded the same, otherwise you would not have put them up twice. Nevertheless, the thoughts, or intents, are redundant. Group them into one thought.

Second, look for things that are similar and could be combined. Combine them. Change the links to show this. Maybe you want to move similar things nearer to each other. It is easy just to erase and do it interactively. No need to keep the prior version as the presentation is evolving.

Third, group related thoughts. In most presentations, there are 3-7 groupings of ideas at a macro level. Draw a border enclosing these similar items, moving around the *Spider* web until the border connects to where you started. It is much like drawing a line of demographics on a map or showing broad geographical terrains such as wetlands or grasslands. Just take all the thoughts that are related to a particular subject and enclose them in a contiguous line. Erase and regroup to make this look as compact as possible.

For example, draw a boundary line around all the items that relate to technology development. Maybe draw a line around all that relate to personnel. Another around all that relate to marketing. Similarly, put a border around all the financial items. These boundary lines weave in and out and wander around so as to enclose related circles. I do not mean to imply that "technology," "personnel," and "financial" are generalized groupings you should use. The enclosed circles are related by whichever criteria you select, and you select whichever criteria are important to the end result, the result you wrote in the center end-results circle.

Finally, number these boundary groupings in order of priority. Which is the absolute, most important thing to get in front of management? The business plan or the additional technology work? The staffing or the capital equipment? Think of prioritization this way. If, three minutes into your presentation to the decision makers, the fire alarm goes off, which one point would you want to have given them before they go flying out the door? This one gets priority number one. This prioritization is not the order you plan to present it, it is the order of importance in getting to the decision, to having the decision makers enact whatever it was you wrote in the first circle.

Which major grouping is absolutely the highest priority? Then, which one is next, and so on. I find that most presentations cannot possibly discuss more than a half dozen items in a scheduled hour's presentation. You should target four. But, at this point, you may have as many as ten or twelve items to prioritize. Let us assume twelve. We will reduce the number later. The first step is to prioritize. Refer to Figure 6-6.

Now, how does one prioritize a large set of actions when each set is complex and there are competing criteria of importance? There are several dozen established techniques for prioritization. They come under the general heading of decision making and include processes such as Decision Matrix Analysis, Quadrant Analysis, Sweet Spot Analysis, Pareto Analysis, Decision Trees, Pros/Cons, Force Field, Cost/Benefit, and Paired Comparison, to name a few. (The names of these techniques are not standardized. They have varying monikers and strongly overlapping functions.) Any one

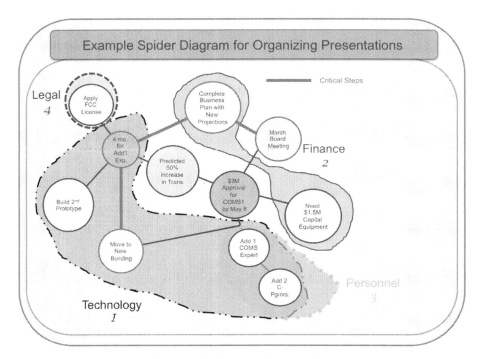

Figure 6-6: The Prioritized Spider

of these techniques and a dozen like them can be used to prioritize your list. A discussion of each would require a much longer and different book. I have used all of them at one time or another. Their formula are analytic. Their outcomes, unfortunately, are heuristic, not empirical. What does that mean? Let us go over the Decision Matrix Analysis and talk about its strength, in theory, and its fallacy, in practice. This will give you an insight.

Decision Matrix Analysis. In this process, one lists all the elements to be prioritized. Then, one lists all the factors that go into the prioritization. In a technical presentation to decision makers, the most important factor is whether its inclusion is required to close the deal. Other factors *might* be how much time it takes to present that element, how difficult the element is to understand, how critical it is to certain audience members, and so forth. Now, set up a two-dimensional matrix. Put the elements in the first column and the factors in the first row. The cross matrix elements then become a relative numerical weighting (1-3, 1-10, or whatever) of that factor for that element. In the end, one adds up all the weights for each element and the numerical values become the ranking. Where's the rub? The numerical weightings. It is easy to make the outcome self-fulfilling. If you are not perspicacious, you ensure the outcome to be your *a priori* prejudice. In this particular example, the best way around this dilemma is to iterate until you feel you have some sense of justice in all the weights. In spite of its shortcomings, I use this technique frequently for close ranking.

Thus, regardless of the technique you use to prioritize, your mind and inclinations are still the deciding factor. The techniques are just ranking tools. Do not be misled into thinking their quantification is measurable. It is not. Here is a technique I use to keep myself honest.

Situational Analysis. Situational Analysis is a string of situations that limit the time available for the presentation. These varying amounts of time help to formulate priorities. You cannot just shorten every element to make them fit. Yet, even if you can, that shortening points out which components you can eliminate as nonessential.

Firing Squad. You are about to be shot by the firing squad (in this case, the decision makers). You have only one thing to say. What is it? That is clearly priority number one.

Elevator Speech. This is the common one that all authors use. You have one minute in an elevator. What do you add that you did not say before the Firing Squad?

Bio-Break. You are off to a bio-break and you encounter the decision maker on the way. You get to go there, wash your hands, and come back. What do you add that you did not say in the Elevator Speech?

Fire Alarm. You start your presentation and then a fire alarm sounds. It is a false alarm. You go outside with the decision maker and then get to come back in about ten minutes. What do you say?

Post Office Queue. My experience is that it takes me about 20 minutes to mail a package at my local post office. You enter the Post Office and find yourself standing next to the decision maker. What do you say?

Dying Words. Most people think of the dying words as what one says during the last gasp. That is not what is meant, here. If you were dying, what would you *not* say? What things are just not worth mentioning? These become your least priority. Do not waste time on these. The Dying Words are the antithesis of the Firing Squad.

These situations just help in calibrating your mind to use the decision-making tools. My *Situational Analysis* is just a tool to calibrate your weightings so that you do not bias your decision.

Experience shows that once you get past item number three or four in priority, it is difficult to put all the others in order. When I confront this, I skip to the bottom and ask the team (or myself, if doing it alone), "Okay, maybe we do not know the priority of items five, six, seven, eight, and so on, but which one, of all the things on our spider grouping, is the *least* important to driving home the end results?" I find most groups easily can identify the least important. Mark that one number twelve, in our example. A team can usually identify the penultimate, and so on for another two or three, in other words the two or three *least* important things. Make these number twelve, eleven, ten, and nine.

At this point, you have the ideas prioritized at both ends, the most important ones, items 1-4, and the least important ones, items 9-12. Do not waste time refining those in between. Just rank those in any arbitrary order to fill in the gap between most important and least important. Do not be undecided and put things side-by-side. Make a decision on every one. You can modify your decision later. Practice the art of becoming a decision maker, yourself.

Learn to prioritize based on importance to the end result.

Now, list the prioritized items separately. This will be the list of topics and their order of importance. Delete the bottom third. They do not make the cut. You are left with nine, which is too long, but you have a logical path from data to conclusions, and a strategy (the prioritization) for getting there.

Now comes the hardest part for a technologist. You have nine major points and need to cut it to four major points. Do it. Reprioritize, group, and eliminate, but do it. Your number one sin as a wizard is trying to stuff too much data and information into the sack. Cut the number down to no more than four. Stick with it. Do not pretend to have four when all you have done is form supersets, that is, trying to look like you have four big groups when, really, you have nine smaller groups. Cut out the superfluous material. If you cannot do it now, you certainly cannot in the meeting. For every hour of presentation, you can hope to cover only four major topics, *at best*.

How does one know when enough is enough and when too much is too much? The number one problem of the technologist is trying to show and explain far too much data in the meeting. Data reduction is initially achieved by following a strategy, discussed above. Further refinement comes by considering how much time you actually have for the presentation. This is covered in Chapter Seven. Final reduction is achieved by more sophisticated means, the subject of Chapter Eight.

MAKING THE FIRST DRAFT

At this point, I do everything else on the computer. Avoid handwritten material as this is difficult to assimilate, transfer, and modify. You have the main points of the presentation. Transfer the *Spider Diagram* into a draft presentation. Compile the draft in viewgraph format, using bullets and phrases. Include everything that you have in the *Spider Diagram*. Roughly sketch in what you want in the way of photographs, graphics, and illustrations. Leave place holders, blank squares or whatever, but make sure that all ideas on the *Spider* appear in the draft.

Make a complete draft of the presentation. (A complete draft must include text, graphics, figures, pictorials, and drawings.)

At this point, some viewgraphs may be simply a blank page with only a title. Figures will likely be blank except for the caption or notes telling what that figure is about, or what goes in it. No matter how rough it is, put together a complete set of draft viewgraphs for the presentation.

Why do this at this point? First, it keeps the end result always in mind: a persuasive presentation. Second, it captures everything on the *Spider*. Third, it keeps the priority you established. (You will find, in working on the draft, that some of your priorities have changed.) Fourth, it lets you build on a skeleton rather than the alternative of having to cobble disjointed fragments and pieces.

Those on the team who are less creative may be capable of doing wonders when iterating, honing, and modifying what others have created. Use each to their best talents. Graphics and data are essential elements. Fill in all the holes and make your first draft as complete as you can. At this point, the draft is not pretty, but it is a complete set based on compelling arguments and reasoned strategy.

SHORE UP DEFICIENCIES

Step back and look at the revised *Spider Diagram*. Are there gaps or holes in it? Are there insufficient data? If so, you may need to put a plan in place to get the additional data or other information. Is the logic correct? If not, refine it. Do the final recommendations flow smoothly from the data and logic, or it a turbulent splash? Is it structured reasoning or hodgepodge wishful thinking? Do your data ignite the recommendations or just sit and smolder? Do the facts form a solid basis or a mound of Jell-O? Are the conclusions transparent or opaque? If any of these are deficient, get started on making them adequate. When you step back and look at it you may find that you are not as confident as you thought you were. If you are not confident, neither will the decision makers be. Get the data you need to make your case. Ensure you have a case.

> Shore up any deficiencies in the data, analyses, or conclusions.

Cancel or reschedule the meeting if the data do not validate your recommendations. Look for alternatives like more time or a revised recommendation. Hopefully, this is a step you will not need to take.

TAILOR YOUR PRESENTATION

Find out who will be in the meeting. The most important thing is to know who will be the key decision maker. Next, which persons constitute a second tier of influence upon the decision maker, and so on. Will there be other knowledgeable technologists in the room? Have outside experts been invited? Who will be at the meeting? The administrative assistants usually have all this information. Get to know them. They are there to help. They want their bosses to look good and they want you to look good in presenting to their bosses. Everyone prefers a happy boss.

> Tailor your presentation and strategy to the specific audience.

If finance is the emphasis of the meeting, then make sure you have given suffi-cient weight to that component. If technology differentiation is the key component, strengthen your presentation to that element. If time to market is key, make sure that is the centerpiece. In other words, tailor your presentation to the interests of the spe-cific audience. If you present the same material in several meetings, to different audi-ences, then tailor to each one. Take the time to do this. Do not be lazy and drift back into the wizard mode of, "Here it is. Take it or leave it." If you want to win, do it right.

Start by getting the job titles of the key individuals who will attend. Know their names, positions, and responsibilities within the company. Ask around to find out how they interface and interact with each other, and how their responsibilities overlap and interplay. The key decision maker may be the final voice that counts, but find out who else might influence that voice. Who does the chief decision maker rely upon? Who does he consult? Who does he ignore? What are his personal interests in this project? What data will he find most convincing?

If it is the first time you have presented to this group, see if you can get a bio-graphical sketch or something similar of the key individuals. For public companies and for government officials, this information is readily available, online, and accessible to you. Sometimes the administrative assistant will send this information to you.

Once you have the information, look for approximate age and gender. Identify the colleges attended and types of degrees for each. Look at prior work experience. As a technologist, I am not an advocate of playing office politics. The purpose of acquiring this information is not to play the game of "colleges and companies." The purpose is to help you tailor your presentation to the specific audience. When you meet with the staff of these individuals, ask questions that will help you prepare. What does the decision maker feel is most important about this issue? What presentations and infor-mation have been given already? This is essential as you do not want to repeat what someone else has done, especially if they were mistakes. You need to know if your presentation will contradict what was presented just last week. What topics should be avoided? Where is the indecision? Where is the controversy? Where are the land mines, that is, the issues to avoid? The staff is usually forthcoming with answers because they want their bosses to be successful.

The intent is not to manipulate the audience, but to formulate your strategy, your logic in presentation, and your reasoning, based on the audience. An effective pre-senter tailors the presentation to the audience.

The Military College teaches courses in military strategy. One, in particular, is the Civil War Battle of Chancellorsville, Virginia. The Confederate Army of General Robert E. Lee faced a Union Army more than twice its size under the command of Major General Joseph Hooker. Lee won the decisive battle (albeit with the regrettable friendly-fire loss of the irreplaceable Stonewall Jackson) not by a larger force, nor by the clever use of artillery, nor by knowing any scientific military techniques. He knew his foe. Lee graduated and trained with his opposing commanders and he knew them well. He knew that Hooker would delay attack until he had everything in place. Lee took advantage of this, by continuing to flank, feign, and outmaneuver until the Union Army was soundly routed. When questioned about his strategy, Lee said, "When they put a man in command whom I do not know, we will have serious difficulty." The tactics Lee used were based on knowing his enemy. The audience of decision makers is not an enemy, usually, but strategy is required to win approval of your recommendations. Lee's strategy was based on knowing his "audience" as should be yours.

The counter-position, that of not knowing your audience, can be deadly. No living individual ever made this mistake twice. No survivor ever required a booster. Here is an example. I was scheduled to brief a senior government official of another state, a state secretary of considerable reputation. She liked our proposal and invited us to be the featured speaker at a seminar she planned to sponsor the following week. I failed to ask the right questions. Huge mistake, especially considering how long I had been doing this. I assumed it would be a cross-section of her department, and it was—except—unbeknownst to our team, she disingenuously invited major competitors to hear the details of what we were presenting, some foisted as consultants. Ten minutes before we walked on stage, one of her subordinates revealed that a major part of the audience would be our competitors.

Now, I had a real problem. If I presented the material we brought, we would reveal proprietary, confidential information that we needed to retain. I next was told that we were the *only* speakers scheduled for this forum. Imagine that. The only speakers. To refuse at this point would be a major embarrassment to the secretary, regardless of her ostensive *faux pas*, or her staff's disingenuous intent. To present our plan would also be an unacceptable disclosure of intellectual property rights. The entire situation was either improper on her part or of one of her staff. It does not matter. The real mistake was mine. The situation would never have occurred had I probed adequately as to the audience. It will never happen to me again. Know your audience.

Developing the Strategy

- Start at the finish. Define success.
- Work backwards to the data.
- Develop a connecting strategy that impels the audience to the conclusion.
- Prioritize the key elements.
- Compile a complete draft of the presentation.
- Recognize and fill in any deficiencies in the data, analyses, formulations, or conclusions.
- Know your audience.
- Tailor the presentation, specifically, to each audience.

Chapter 7: Speed Limits
Managing Time

Speed is a great asset; but it's greater when it's combined with quickness—
and there's a big difference.
— Ty Cobb

ony Hillerman writes mystery novels whose protagonist is Navajo law enforcement and whose setting is an Arizona reservation. The plots sometimes evoke "Navajo time;" if a Navajo arrives within an hour of the scheduled time, that is considered "on time." It is akin to a reply at the Mumbai, India railroad station. Anxiously looking for an already tardy train, I asked a nearby man, a local Hindu, if he thought the train might arrive within the next ten minutes or so. He admonished, "Perhaps. But, if not today, then tomorrow, maybe."

In a colonial home in Boston, I encountered a mantle clock built without a minute hand. The guide explained that some clocks were made with an hour hand only. I looked at the hour hand and could see that it was approximately halfway between the hours, but there was no minute hand to tell precisely the fraction. And there was no GPS to keep it in synchronization. The guide explained that when that clock was built, being fifteen minutes late was considered being "on time" because of the huge variations in the actual ability to know the correct time.

One of the major problems of 15th-century sea navigation was the absence of an accurate timepiece. This lack made the sea voyages of Christopher Columbus and Ferdinand Magellan mostly a day-to-day guess. In 1761, Great Britain paid a prize of 20,000 pounds, equivalent to several million dollars today, for the invention of an "accurate" clock.

Even inexpensive wristwatches, today, are accurate to a few seconds per month. Add a GPS synchronization, or a few dollars for more expensive technologies, and the accuracy overwhelms. Does this give you any indication of how important time is to those who measure seconds like specks of gold dust? They are not wizard scientists. They are business and political decision makers. Their time is valuable, at least to them. It is imperative that a technologist know how to tell "business time."

First, the no-brainer. Arrive and be available not less than 15 to 20 minutes before the scheduled time. That does not mean driving up in the parking lot. It means being there in case you are requested to begin early. You have a timepiece. Use it. Call ahead to see if the meeting is running early. If so, adjust accordingly. There is no excuse for being late, none. However, getting to the meeting on time, albeit essential, does not merit a chapter. The broader question does.

Most technologists can get to the meeting on time, but once they step to the platform, they shift to Navajo time. If they glance at the clock at all, they squint through a glass, darkly. The wizard runs long, frequently unbearably long, often intolerably long. This leads to frustration on the part of the decision makers and failure on the part of the wizard.

Time is not your friend. There is a current business maxim, "early is on time, on time is late." They possess strong sentiments regarding time. However, they are not talking about getting to a meeting on time, they are talking about delivery of product.

Let us discuss then the delivery of the product. Go back to the moment you first learned about the meeting. Assume that your presentation was scheduled for one hour, starting at 10:00 AM on a given date. Should you prepare sufficient material to speak for 60 minutes? No. Well, how about 50 minutes? No. How about 45 minutes? No. Well, then, if a technology presentation is scheduled for one hour, for how much time should one prepare? Forty minutes. (If you are scheduled to present for half an hour, then prepare twenty minutes.)

STARTING WITHOUT THE KEY

First, some lessons on how to tell "business time." Regardless of the clock, the meeting will not truly "start" until the key decision maker arrives. I know senior leaders who are martinets about time. They start and end every meeting to the minute. I also know executives who never once have started on time. Most are, expectedly, in between. The typical scenario is that the key decision maker's prior engagement will run over time and cut into your hour. Your own meeting may appear to start but, regardless of any discussions that may occur before the key decision maker arrives, your presentation does not really start until that person is seated and ready to listen. When the administrative assistant tells everyone, "The president said to go ahead and start, and that he would join you shortly," you know you have a problem. "Shortly" means "indefinitely," in business jargon.

In such case where you must start the presentation without the key decision maker, begin the presentation as you would normally. When the key decision maker does enter the room and is seated, review what you just told everyone else in the room while the key person was gone quickly, and without any formality or restatement of purpose,. Do this even if the key individual says, "Don't start all over, I'll catch up." But do *not* make it obvious that this is what you are doing. If you announce that you will start over, you will be told to just go ahead; so do not announce it. You need the key decision maker to be on the same page as the rest of the audience, so do a quick review without making it look like a review and certainly without saying so. Make it succinct, but do it. It never hurts to reinforce your points, anyway, so emphasizing them is not detrimental. Use different words and make it flow seamlessly. Definitely do not just backtrack through the same visual aids. The finesse is in iterating without repeating. Take only a few minutes, though, to bring the key decision maker up to speed. If it is noticed that you are repeating or that you are taking too long, do, indeed, skip the review and move on with the presentation as requested. You will only get one chance to bring the key decision maker up to speed. Practice ahead of time how you might do this in case the situation arises.

HOW LONG DO I HAVE?

Maybe you are lucky and the meeting does start on time with all the right people seated and in the room. What is the problem with time, now? The problem is that the attendees will want to make it to their next meeting on time, whether you start late or not. At least one of the key decision makers, if not all, will have a meeting scheduled for 11:00 AM, supposedly your end time. Even if the key person was ten minutes tardy arriving at your meeting, thereby forcing your meeting to start late, it does not matter. All of the decision makers will have something planned (formally or informally) for 11:00 and all of them will have calls, emails, or restroom breaks that they must make in between your meeting and the one at 11:00. If they will need to be in another room by 11:00, they require transition time so they can start that meeting as scheduled. It is common for a wizard to think that if the meeting starts late, the decision makers will give her extra time. This is erroneous and a clear sign that you have not yet learned to tell business time.

Add to these initial and ending time losses the hopeful premise that the audience will be interested enough to ask questions. If they do *not* ask questions, you have an even greater problem: either they did not understand what you presented, or they did not care, or both. Of this, more later. The more important or controversial your topic, the more questions should be asked. The better you become as a presenter, the more you will pull them into your presentation to interactively ask questions that will engage them and make your job easier. This is much preferred from a presenter's viewpoint than asking them to save questions until the end, which is mostly ineffective. You must allow time for questions and discussions.

Finally, there is another thing that takes up your time. You are there to persuade the decision makers to do something. So, when they decide to do it, specifying the implementation and documenting an action plan takes a little more of your scheduled time.

For a one-hour presentation, you should plan on no more than 40 minutes worth of material that you actually present. On very important or very controversial topics, a *planned* 30 minute presentation (planned by the wizard) can easily consume an hour when you include all the time delays before and after, and all the questions, discussions, and actions.

Ignore this sage analysis of time only at considerable risk. Wizards of all disciplines and all levels of seniority persistently think they are unique. They believe, every time, that they can stuff 10 pounds of presentation into the 5-pound bag. The wizard thinks, "In my next presentation, they will give me more time. In my next presentation, I will move through the material faster. In my next presentation, I will be able to explain it better because they already know about it." You are falling into a wizard trap. It will not be. Make no mistake, a 40 minute *actual* presentation is maximum, absolutely maximum, for a *scheduled* one-hour technology presentation. If you happen to be the one-in-a-million miracle child who actually finishes, with questions and approval, in 40 minutes, if Moses raises his hands and causes time to stand still, the audience will only love you the more for giving them back some of their valuable time.

On more than one occasion, I have been scheduled for one hour and, because of conflicts with the schedules of the key attendees, and other issues, in the end I had less than 20 minutes total to give the entire presentation. The choice is then whether to come back at a later time or do what one can with the time available. High-level meetings such as with the president and his entire staff, with entrepreneurs, or similar require advance scheduling; a postponement can be weeks or months into the future. By the time the meeting is rescheduled, your window of opportunity may be gone. We will discuss in detail later how to prepare for a massively truncated time period, but suffice it here to say two things. One, you must compose your written and visual material to stand alone so you can leave the complete information with them. Two, you must have a contingency presentation practiced and ready for the shorter time period.

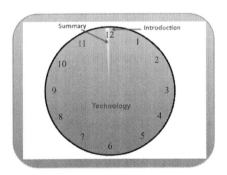

WRONG: *If the meeting is scheduled for one hour, most wizards feel they have nearly a full 60-minutes for their presentation. A minute or two to start, a minute or two to end, and the rest of the time is presentation.*

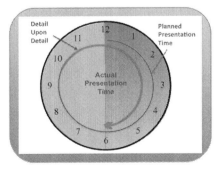

WRONG: *This is what usually happens if a technology wizard is scheduled for one hour. The wizard plans for a full 60-minutes of presentation and then takes an additional 30-minutes for details.*

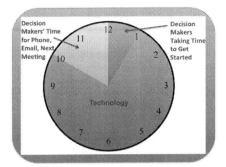

EXPECTED: *This is what the decision maker expects—about five minutes to get started and about ten minutes at the end to make phone calls, email, and get to the next meeting.*

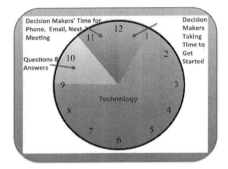

IDEAL: *This is success for the wizard. Satisfy the decision maker's expectations and still allow time for questions, answers, and actions.*

If her prior meeting ran late, a decision maker may tell you to "take all the time you need," as an apology for her starting late. Do not mistake this sense of courtesy for something it is not. Remember, the goal is to *persuade*, not to finish the presentation. (Chapter Five's Brer Rabbit and the briar patch?) That executive is still on a schedule of some sort. Once you go past the originally allotted time, her mind will shift elsewhere and the opportunity to persuade will fade exponentially. You do not have more time.

SHORTER TAKES LONGER

Whatever the topic, putting a good technology presentation into just 30-40 minutes is a major challenge. It takes hours of effort to hone the data and formulate compelling and persuasive arguments. The apostle Paul opined, "I would write a shorter letter, but I do not have the time." The shorter you must make the presentation, the more time it requires to prepare it.

If the meeting is running behind schedule, as most will, you cannot gain time by speaking faster and rushing through the material. The pace of the presentation is driven by the complexity of the subject matter and the cognitive knowledge of the audience. The wizard must make the presentation fit the time without rushing through key items. And the time given changes dynamically as we have discussed. By rushing through everything, you may finish the presentation on time, yes, but not achieve the results desired. What good is that? It will be a pyrrhic victory, finishing, if the decision makers do not approve and enact your recommendations. Rushing also accentuates any apparent (or real) perception of nervousness or stage fright and takes away from the professionalism of the speaker. Therefore, you must condense and truncate the content, *not* try to speed vocalization.

If the dynamic time allocated is very short, you must compress some items and leave others out entirely. If you planned for 40 minutes and when you come to the meeting, it starts late and you have only about 30 minutes of actual presentation time left, you cannot simply *compress* everything into the smaller time because then nothing will be understandable. You must leave something out. You can, and sometimes must, do this omission in real time, but it is far better to plan on this contingency and prepare for such an event ahead of time. In some sense, it is like military strategy and one must prepare for contingencies. Therefore, when you review your 40-minute presentation, as discussed prior, go through the steps of how you would compress it to 30 minutes, should that become a requirement. Try to do the same for 20 minutes. For 10 minutes. For 3 minutes. Compare this to the *Situational Analysis* method we discussed in Chapter Six. This reduced-time exercise is not in vain. It makes you, the presenter, get foremost in your mind what is important, what is key, and what is essential. You will discover almost every time that some items you felt just could not be left out were, in actuality, not essential after all. You may decide to simply discard them because you discover them to be peripheral and not central to your persuasive argument.

This practice at compression hones your skills as a presenter and a technologist. It makes you read your graphics (visual aids) and text for those truly essential elements. Often, you will find yourself wording and rewording, to get those key thoughts just right. By all means, do this. Never mind that the presentation has gotten shorter. Your expanded knowledge as a technologist and presenter will fill the gaps and make it fit the time like a glove. A word fitly chosen is the best way to get your points across.

BREAKING THE 20 MINUTE ATTENTION SPAN BARRIER

There is a widely accepted rule, outside the business world, that a speaker can hold an audience's attention only for about 20 minutes, so one should limit presentations to less than that. This twenty minute attention limit possibly is the Gaussian mean for the attention of general audiences like parishioners absorbing a sermon, parents at a school meeting, or diners enjoying an after-dinner speaker, but not so for an audience of decision makers.

So, do decision makers have a longer attention span than, say, 20 minutes? No. Much shorter. Decision makers have an attention span challenging that of a 3-year-old without lunch. Decision makers may *sit* in your meeting for an hour, but *keeping their attention* is a continuous challenge. To keep their attention, you have to keep the presentation dynamic. That is, dynamic in their world. You must keep putting decisions in front of them. If you keep challenging them, their short attention spans can be stitched into 40, 60, or even 90 minutes. If they are not challenged, they will be out the door, always mentally and sometimes physically, before you even get started.

The mismatch is that decision makers have a multiplexed, microsecond attention span, but the technology is so complex it cannot be explained in less than 40 minutes of presentation time, and even that is a challenge. How can a wizard keep decision makers riveted for an entire presentation?

The technology presentation that persuades is a series of strategic elements. The trick is to meld these elements into a cohesive whole and lead up to a decision in your favor, like a spellbinding novel reaching a climax. The segments must be logical threads that weave a fabric of decision. Thus, it is not like you really have 40 minutes to say what you want and they just listen. You must make sure that every 2-3 minutes you are challenging them with information, decisions, or items of interest to them. This is a colossal challenge to the wizard presenter, because wizards are accustomed to a much slower pace of elucidation.

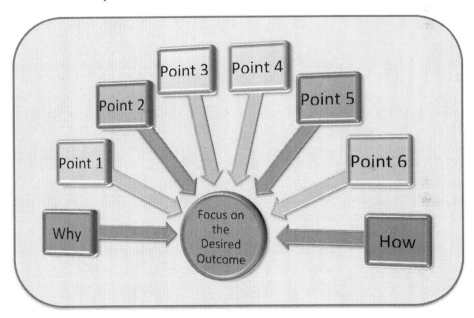

Figure 7-1: Every Element of the Presentation Should Focus on the End Results

PACING THE PRESENTATION

Now that we understand why we must prepare only 40 minutes of material for a 60-minute meeting, how should that time be allocated? In other words, how much time should one spend on each chart, each viewgraph, each visual aid or "slide," so to speak? Ask most managers and you will get the rule of thumb, "You should spend no more than three minutes on each viewgraph. You have fifty minutes (incorrect) to present, so figure three minutes a viewgraph. That's 16 viewgraphs plus a title view-graph." Give most wizards this direction and they will come to the meeting with 28 viewgraphs and 10 backup viewgraphs. They will take ten minutes on each of the first three and burn the time up in the introduction. No wonder wizards have such a bad reputation as presenters.

A properly designed presentation does not travel at a constant pace. The speed is dynamic. It is 30 seconds on this viewgraph and two minutes on that one. Five minutes is an absolute maximum to spend on any one viewgraph, and that had better be a *Visual Aide*, which is a special product we discuss later. The pace is dynamic. Laying the foundation requires more time than reciting status, for example. A technology presentation, then, is more like a series of sprints, whereas most wizards treat it as a marathon.

TIME ALLOCATION

The guideline of three minutes per viewgraph is not bad for a first approximation. Still, you must stick to this plan to make it work—most wizard presenters cannot. Personally, I find that using three minutes per viewgraph makes the presentation too slow. Clearly, you lose your audience if you spend more than three minutes and less is usually better. I generally plan on talking two minutes per viewgraph and, in preparation for the meeting, adjust the content of each viewgraph accordingly to achieve this faster pace. But this level of skill takes a lot of practice and strong discipline. You, the presenter, must meter the flux of viewgraphs and mentally keep track of the clock and where you are in the presentation. If you can maintain this pace, then the decision makers have no free time for wandering thoughts. It is not a matter of speed, it is a matter of flux. How much *cogent* information can you present per minute? This depends on how much time you spent in preparation (longer makes shorter), how skilled you are in delivery, and how complex the subject, the latter actually being the much lesser effect.

The best approach is to take the time you need to adequately explain the subject to your audience. Then, adjust each viewgraph so that each one takes only about two minutes. It requires a great deal of work ahead of time to get this right. Obviously, I am speaking in this chapter about viewgraphs of content. Title viewgraphs and similar are not even counted.

From a practical viewpoint, then, you must have a clock readily visible to you during the entire presentation. If there is not one in the room, and this is common, carry a small one and put it on the table in front of you where only you can see it. You will not need to *look* at it constantly, but you will need to be constantly aware of the time. Looking at a wrist watch is distracting. Do not do this. It is far preferable to have one of those little travel clocks facing you. Choose a small, black one that will not draw attention to itself. If you forget, take your watch off and place it somewhere where you can see it unobtrusively.

I have given presentations that averaged 70-90 seconds per viewgraph, but those presentations had photographs, graphics shown in layers, and other factors that would make the pace *per viewgraph* seem faster than it actually was. Also, these were standard presentations that I gave dozens of times and I had a professional technician advancing the viewgraphs from my hidden cues who was himself, keeping track of time. Without these additional factors, I would have managed only two minutes per viewgraph, which is my recommendation for seasoned presenters and a goal for every technologist. But remember, if you cannot stick to your own time limit, then you have only fabricated an excuse for presenting more material than is needed and taking forever to present the subject. Do not target the clock, target the flux, that is, the amount of information you can convey for a given amount of time.

In managing the clock, discipline is everything.

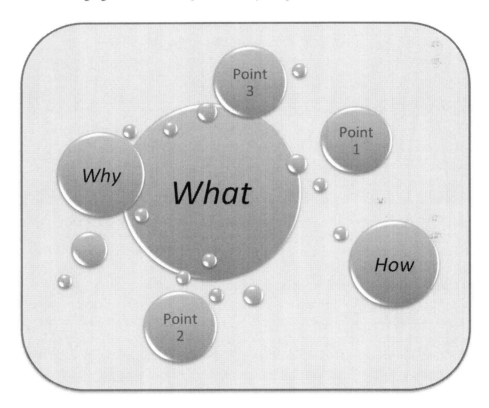

Figure 7-2: Dynamically allocate the time for each viewgraph but be conscious always of the total time.

BAD NEWS IS A TURTLE

Good news travels at the speed of light. Bad news drags on. Vacationing in France, we boarded the train at Calais bound for Paris. The plan was to stop at Beauvais and then travel the additional 55 miles on to Paris. Beauvais is the city with the infamous *Cathédrale Saint-Pierre de Beauvais*. This wonderful, Gothic cathedral is an engineer's nightmare of flying-buttress architecture, built and rebuilt over eight centuries. So, we board the train for Paris, wanting to stop en route at Beauvais. The dapper porter with his little pillbox hat did not *speak* English, but he *understood* my question of whether I should take the express train into Paris and return to Beauvais, or take the local train to Paris and stop at Beauvais along the way. "No, no, no," he fans his index finger like a vertical pendulum. "Choo, choo, choo," he says, and he makes his fingers into a little train that he jerks along in front of his vest, indicating that the local train stops everywhere and will take hours of additional time.

Bad news is a local train. It stops everywhere. Every member of the audience has a question, wants to know why, wants to know how, wants to know which head to lop off and who to punish. Bad news breeds scapegoats, assembles witch hunts, and reserves guillotines. Adjust your presentation time, accordingly. "Choo, choo, choo."

KNOW WHEN TO QUIT

You know the aphorism, "winners never quit?" Well do not say that within earshot of a wizard because wizards never know when to quit. They keep going, and going, and going. They cannot read an audience. It takes observation to know when to quit. To the wizard, this is difficult. Work to acquire this skill. I do not mean knowing that the time is up, but knowing when to quit, otherwise.

First, let us address the positive outcome. In basketball, once the buzzer-beater swishes through the hoop, put the winning points on the scoreboard and head for the locker. Do not continue to lob air balls. It went through the basket, so move on. In baseball, when the ball goes over the center field fence, and the crowd leaps to their feet, then you run the bases and keep going to the dugout. Do not stand there gawking (and talking), waiting for the next pitch. In football, if you get the ball over the goal line for the final touchdown, you dance to the sidelines. The game is over. In a technology presentation, once you make the sale, once you get agreement, once the Grand Poobah of decision making nods his head, then jump on your horse and high-tail it out of Dodge. In other words, if you are asking for money for a major research project and the decision makers agree to fund it, then it is time to close out the presentation and go count dollars. If you ask for a promotion and get confirmation or you request more personnel and get agreement, it is time to quit. So, quit. They might change their minds if you keep talking, so do not.

On the other hand, if you wake up in the middle of the presentation to find yourself in *Madame Tussaud's Wax Museum*, you have a problem. If you see eyeballs rolling across the floor, or whiplash and concussions, you have lost your decision makers.

They are somewhere else. What do you do when you wake up to realize you have lost them, when you realize that their sole purpose now in life is for you to quit? You can try some techniques, discussed later, to bring the mannequins back to life, but if those techniques do not resuscitate the decision makers quickly, it is time to quit, regroup and try another day. Continuing to drain their dead batteries will only make it more difficult for you to return. Ask them what they would like to see, how you can prepare differently, and then quit and do just that. But, by all means, quit.

It may not be your fault. The decision makers may have just learned that they did not make their financial goals for the quarter. Or, it was just announced to them that they will not receive their annual bonus. (No way to wring success out of that.) They may have been notified only moments before that they lost a "must-win" contract. Maybe the FDA has turned down the new drug. Maybe the company just announced a recall and they anticipate millions in fines. There can be many reasons that do not involve you, but do not add to their woes. Quit and return when they are responsive. There is no use presenting technology when no favorable decision is possible.

Most often, however, a dead audience means you killed them, probably with all those needless technical points. Death by a thousand details. You prepared the wrong material. You got off on too many tangents. The data are insufficient to warrant the meeting in the first place. The timing is not right. Whatever the reason, go fix it, but value their time, and yours, and know when to quit.

This "knowing when to quit" is not innate in technologists. It is acquired only by force feeding. My company was working on a major proposal for a Middle Eastern country. This proposal was phase one of a project that would last ten years and total over a billion dollars. This contract went through the US government, as is common, and the bids and proposals were governed by US law. We traveled to the Middle East a number of times. We knew the foreign and US command from tip to toe, knew them all personally. Our technology solution was the one favored by the customer. We were miles ahead of the competition. Miles.

This six week long effort was due to culminate only a few days before New Year's. All bemoaned the effect it had on their families. The proposal team worked through Hanukkah and Christmas (not uncommon). The finished proposal was **due in Washington, D.C., at the government contracting office, 9:00 A.M. Eastern Time, on Thursday of that week.**

There is no such thing as an 8-hour shift when you work a proposal. You work sixteen to eighteen hours at a stretch. This team of about twenty-five people was not shy about putting in the effort. They finished the proposal at 11:00 P.M., Tuesday night. All that was left to do was edit the copy, print, and ship. The team went home for a well-deserved holiday. The proposal manager, his work finished except for final details, left for a previously planned cruise (plane tickets, boat reservations, the works). The vice president of the division remained in charge of the final copy-editing, printing, and shipping.

Early Wednesday morning, the proposal went to the print shop. Editing the final copy, printing, and packaging would take about six hours. It was to be shipped Wednesday afternoon on commercial airlines. There were a half dozen commercial flights available on different airlines so plenty of backup. Duplicate copies would be sent by on two different flights, one in the early afternoon, one early evening. Both shipments would arrive by Wednesday night in Washington, D.C. Company employees already were stationed in Washington, D.C. waiting for the boxes to arrive at Reagan National Airport. Those employees would pick up the boxes in their personal cars and take the proposal to the contracting office to be waiting when it opened at 8:00 A.M., Thursday morning. Backup employees were on standby in case they were needed. The weather was beautiful at both ends.

This was a proposal hundreds of pages in length, multiple copies, with precise, complicated formatting and page limitations for all sections of the text and types of illustrations. In proposals like this, the number of permitted pages is short (remember, shorter takes longer) so editing is a critical job. Sometimes, you must cut out a sentence or shorten it to fit the final print format. (Commercial printing machines are not identically compatible with all available computer text and graphics software. "What you see" is not always "what you get.") At about 2:00 P.M. on Wednesday, the vice president, seeing that they were almost ready to print and ship, left to attend his son's holiday event that afternoon, and a city board meeting that evening (he was an alderman). He left the assistant proposal manager in charge of the team for final editing the final copy, printing, and delivery to the airport.

Now the assistant proposal manager was forty-two years old, had a Ph.D. in physics, and had been with the company twelve years. He had spent all his career in technology organizations and only recently moved into management. This was no novice wizard. He was a professional wizard gone over to the other side, management.

At 11:00 P.M., Wednesday night, the manager of the print shop calls the vice president at his home (yes, he was asleep) to tell him that "maybe" there is a problem. The assistant proposal manager is still making changes to the proposal. It is not printed. It has not shipped. (The last commercial flight departed two hours prior.)

"What?" the vice president screams into the phone. (And more.) "Stop, what you are doing! Immediately! No matter where you are, stop! Print whatever you have and box it up!" He tells them he is "on his way." The vice president makes the twenty minute drive to the facility and finds the printing and binding almost complete by then. The assistant proposal manager is yet sitting at his piled-up desk in the print shop, with his small team, still making changes, and still thinking there is time to incorporate them. He is ignorant of the gravity of the contract specifications, oblivious to the time, and totally above the complex logistics. (He is a *technology wizard*. He is still editing and making changes, you see.) At midnight, the vice-president charters a private jet. (Money talks and cost is no object at this point.) It will take several hours to get the pilots (two are required), get the airplane ready, and set the flight plans in place.

The drivers break all speed records getting the boxes of proposals (multiple copies of multiple volumes) to the airport. Same on the other end in Washington. By 8:00 A.M, Eastern time, the plane is making its approach to Reagan National. The time is critically close, though, because the plane has to land, taxi, unload the boxes, and the drivers must navigate traffic to get to the contracting office in the suburbs. The vice president calls up the colonel, who was waiting at the contracting office, and says, "Can you give me some slack. We have a potential problem." The Colonel, knowing we had the solution he needed and wanted, says, "I'll see what I can do." A short time later, the car screams into the lane, thee of our employees jump out and manhandle the boxes into the contracting office. The clock reads 9:23 A.M.

Now what?

The contracting officer looks at the boxes. He records the time. The boxes sit. No one will touch them. They are cordoned off like little lepers. People in the office walk around them and wag their heads. Competitors snicker. The winning proposals lie stillborn in their cardboard wombs. The contracting officer declares a default. My company files suit. Competitors file a counter suit. Within a few weeks a competitor is declared the winner. No one opens those boxes. No one reads a single word of the "winning" proposal. The pallet is shrink-wrapped and shoved somewhere in the bowels of US intelligence. On the final reckoning day, they will be expunged to that great shredder in the sky. For all that effort, the company never earned a dollar.

What happened back at the ranch? There were not enough fingers to point nor directions to point them. Executives queued up for a quarter mile wanting their turn to operate the guillotine. Who was at fault? Management, of course, failed in its checks and balances. Yet, the firing pin for this grenade was placing a technologist in charge who had no recognition of when to quit, no knowledge of why he should, and no ability to do so, regardless. I am sure that every correction he made was an improvement. I am sure that every word he wrote was better than what existed. I am equally certain that had he not been stopped, he would be sitting there, today, still making changes, still making it perfect.

THE FAT LADY WAS NOT A WIZARD

Listen for the Fat Lady. When she sings, bring down the curtain, with or without applause. Most technology presenters do not even know what the Fat Lady looks or sounds like. Learn to recognize her. For instance, if you request that a new project be started and the decision makers say, "no," then, that's the Fat Lady singing. Do not waste your time and frustrate them. At this point, accept the decision, find out if there is opportunity for the phoenix to arise and, if so, find out what information they might need to reincarnate him. This is far preferable to continuing to ask for something against which they have decided already. Do not be deceived by tact or civility. If you ask for a rematch, they may politely say, "yes," but the meeting may never occur or, if it does, it becomes only a formality. Learn to listen to the Fat Lady. If she sings, it is over. You must know when to quit.

–Time–
The
Crucial
Element

o Plan a presentation 2/3 as long as the time allotted.

o Plan, specifically, for compression and deletion.

o Practice your "elevator speech."

o Pace your presentation, but remember that the pace is dynamic and depends upon the material and the audience.

o Bad news takes much longer to present than good news.

o Always be aware of the clock and where you are in your presentation.

o Know when to quit; know when you win; know when you lose.

o Thou shalt not kill.

Chapter 8: Formulation
Distilling the Concepts

Experts often possess more data than judgment.
— Colin Powell

O n a hot summer day, the temperature at Hoover Dam can reach 115° F. On one side is massive Lake Mead. 725 feet lower, the other side is the foaming Colorado River, bursting through sixteen turbines from a pressure of 45,000 pounds per square foot. From these turbines, eighteen hundred megawatts of electricity surge toward Las Vegas. Many wizard technologists operate like Hoover Dam.

The wizard walks into a presentation carrying tons of data and information and knowledge in his head. It is a massive, comforting reservoir. He faces the decision makers. The wizard reservoir of knowledge spins the intellectual turbines and the presentation proceeds at full bore. The turbines are whirling and decision makers ignite to hear about technology and what it can do for them. But something goes wrong. Uncertainty sets in, an unexpected question is asked, or the wizard forgets why he is there. The wizard-temperature rises and the wizard-DNA gets, well, wizardy. Instead of the turbines churning and the decision makers getting all fired up, the water is uselessly gushing out, surging, and seething. The wizard-adrenaline opens the valves to increase the flow of data, (they must need more data!) but the turbines are overwhelmed. Water pours over the spillway. The decision makers drown. The wizard's favorite project washes away in front of him, doomed to a watery grave.

The absolute number one sin, the number one error with technology wizards, is presenting far too much technical data and information. It is data and information that no one else understands and no one else wants to understand.

Our team was briefing ranchers in Montana about a system for tracking cattle. The presentation was too technical and too detailed for the audience. A rancher bemoaned with the old saw, "When I go to feed my cows, I don't load up all the hay I've stored in the barn. My gosh, I feed them only what they can eat."

Even Goldilocks knew that the bed should neither be too long, nor too short, but had to be "just right."

LEARNING TO BE SELECTIVE

A presentation overburdened with facts, figures, and data may seem perfect to the scientist, but to the decision maker all those details are just amalgamated confusion.

> *Pare and compress the draft to fit the amount of time you will present.*

And, the least developed skill of any wizard is knowing how much is enough.

While the vast majority of technologists sin by giving far too much data, I have seen a few so arrogant as to go the other way. They present little real information, thinking that only the village idiot would dispute their expert opinion. Then, they are amazed to observe decision makers drawing back in skepticism—their opinion, apparently, not sufficient.

There is that holy grail of technical presentation: just right. Just the right data, just the right amount of data, just the right logic to the data. Just enough information. You can love it when a plan comes together. But, this exactness, this saying and doing just the right thing in the right amounts, is difficult. If I had to find the greatest weakness in technology presentations, this would be it. Wizards have no skill in being selective.

Presenting technology is not a card trick. You do not arrive at persuasion by fanning the entire deck and hoping the decision maker randomly draws the right card. You get to the "just right" results by designing, crafting, and molding a presentation for the specific audience and the subject. The presentation that is perfect for one audience may be inadequate or inappropriate for another, even if the subject is the same. Your goal is the correct amount of data, the exact depth of information, and an efficacious delivery for each unique audience. How do you ensure that the presentation is "just right?"

First, you must carry into the meeting considerably more than you will deliver. You will put only five pounds into the five-pound bag, but you must carry twenty-five pounds in your brain. The presenter must have more, much more, in the "hip pocket," so to speak: more information than is ever shown, presented, or formally prepared. But it is not just random knowledge. You must possess quantified, current, targeted facts apropos to the specific subject.

Consider an opera. The vocalist performs an aria, mellifluously accurate to every note of the composer. But, what if the audience demands an encore? Does she not have more? Consider also a war. We need artillery sufficient for the enemy. We can plan. We can strategize. But when the shooting starts, we need to have a ready arsenal in case the enemy is greater or the weapons different than we expected. Our little horse cannot be just a one-trick pony.

The same with a technology presentation. The presenter must have a depth of knowledge far in excess of the actual material presented. This extra depth and breadth interjects complementary material spontaneously, as the speaker instinctively reads the reaction of the audience. This additional complementary body of knowledge reinforces, shores up, aids, and navigates the presentation in real time. The prepared presentation is like a boat. The boat floats all right, but the additional knowledge adds sails, masts, beams, and rudder to transport the ship across whatever waters it encounters. You do not know what the waters will be like until you get into the meeting. No amount of practice, no amount of training can compensate for lack of subject knowledge. But you do not dump the entire load. You select.

If a swimmer in dangerous waters is attacked by one shark, the other sharks smell blood, go mad, jerking into a frenzy, advancing, attacking, and churning for prey. Human nature reacts similarly. Either consciously or subconsciously, once the audience realizes that the speaker is out of her region of knowledge, they attack. Sometimes consciously challenging, sometimes silently skeptical, and sometimes even pitying, they advance. Once they smell blood, they all charge. But let that same speaker counter with succinct, right-on-the-mark answers, and the rough waters will calm. The sharks will move on. The presentation will sail smoothly. Let that speaker fill in with cogent material, or draw from complementary data in her personal knowledge base, and the audience will soon be giving the presentation for her.

This "domain knowledge," as it is called, is critical. My worst nightmare is when I must present someone else's material regarding a subject in which I have limited first-hand knowledge. (People get ill, miss planes. It happens.) The set of questions I can answer is limited. Take the required time to study, investigate, ask questions, and gain sufficient domain knowledge to be the expert, at least in the presentation room.

In a technology presentation, one way to get quickly into trouble is to wander outside one's knowledge domain. Like Dirty Harry, "A man's got to know his limitations." (It applies to both genders.) You must know what you do not know. Do not step out into waters that are over your head. Stay within your domain of expertise. It is not necessary, nor desired, that you tell the audience all the things you do not know, but it is necessary, and required, that you know what you do not know.

Do not give answers in ignorance. Better to say, "I don't know," than to pretend to know and give a wrong, misleading, or inadequate answer. The best response is to know, of course, but that is not always possible.

We were in New Orleans, my son and I. We adventured to embark on *Dr. Wagner's Honey Island Swamp Tour*. It was in the late afternoon when we arrived and the tour was to last two hours. We took off from Dr. Wagner's dock and motored miles into the swamp, going this way and that, circling who knew where. We looked for alligators and snakes and other creatures of the swamp. About two hours passed and it began to get dark. The driver of the boat stopped and cut off the engine. I looked about. It looked different in every direction and it looked the same—cypress and moss, cypress and moss, everywhere, swamp beyond swamp, fog and mist, shadow and light, distinct without distinction. The sky reflected in the water, the water reflected in the sky, everywhere the same and everywhere different. It was eerie. Our guide informed us it was not uncommon for a sportsman to launch out for an evening's fishing, get lost in the swamp, and die of exposure before he could find his way back. I believed it.

Stay out of unknown waters. Make every attempt to keep the audience in waters familiar to you. Always keep the landmarks in sight and always point them out to your audience. If the audience gets lost in all the data, you cannot lead them out.

Do not step out of the boat. Most wizards cannot walk on water. There are gators in those swamps and the banks are full of quicksand. Stick to your strategy and your plan. Use the advice of Chuck Noll, who has won more Super Bowl rings than any other head coach. "Leaving the game plan is a sign of panic, and panic is not in our game plan." Stay on task.

Hopefully, you will be presenting your own material or material with which you have good familiarity. In which case, you have little to worry about in the way of technology knowledge. But it is always a good idea to review even familiar material a few days before the meeting and again just a few hours prior. Think of possible questions that might be asked and then formulate an answer. If you do not know the answer, find out. Be as quantitative as possible. This process of asking and answering questions has a secondary effect. Questions can reveal gaping holes in your data or logic that may need repairing. Do not wait until the middle of the presentation to find them. Give yourself time to plug up those holes that would sink your boat. Better to do that before the presentation than drown in front of the decision makers. Tossing life preservers is not one of their skills.

Most of the things you learn as backup and additional information will not be needed in the presentation. But, you do not know which ones are needed and which ones are not needed. There is nothing wrong with having extra ammunition, just do not shoot every round in the clip. What a wonderful feeling it is to have the right answer to a spontaneous question from the audience. No hemming and hawing, just the right answer. But do not keep firing.

Again, if you do not need the information, do not bring it out, do not mention it, do not show it. To do any of those is to invite failure. Use your vast head of knowledge for receiving and assimilating, not broadcasting.

VISUAL *AIDES*

With all that good data to show, it is commonplace for wizards to cram it all on one viewgraph using 3-point font and no margins. The typical wizard points to such a viewgraph and says, "It's all there on the viewgraph. You have a copy in front of you if you want the details." This produces two undesirable thoughts in the minds of the decision makers. They think either, "If the details are *important*, why didn't you make it so I could read them?" Or, "If the details are *unimportant*, why did you show them to me in the first place?" You cannot have it both ways. If it is important, make it readable. If it is not (which is usually the case) then do not show it. Or, at least do not show it in that form.

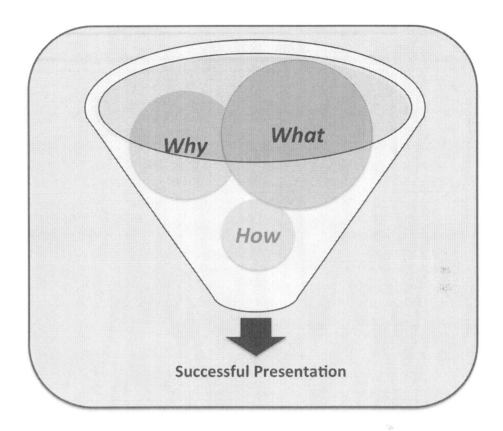

Figure 8-1: Funnel All Actions. Use only those that achieve the end result.

Here is how to handle all that detailed data. College communications courses correctly stress the importance of a visual aid, *per se*, but not in the way required for technology presentations. In the communications course, a visual aid can be a photograph, a drawing, something you hold, something you point to, whatever. So, in a typical *public speaking* scenario, the visual aid can add humor, be a transition to the next topic, or just be illustrative. In a speech, the visual aid most often illustrates a point and keeps the attention. For example, if you were telling about your trip to Granada, Spain, you might show pictures of the Lion Fountain at the Alhambra, snapshots of the craftsmen inlaying intricate wood patterns, show the streets of the city, or maps of the route. But this is not usually of much impact in a technology presentation to persuade decision makers. You are not making a *speech*. True, you may have pictures of your laboratory, or the new product, but these are just illustrative, not persuasive.

Hence, I want to teach a technique that I call the Visual *Aide* (not *aid,* catch the homonym). A Visual *Aide* is a viewgraph that helps you explain ideas; it is an assistant to you, it is not simply illustrative. Think of it as artillery blasting through those cerebral barriers of the decision makers—inattention, misconception, preconception, fuzziness, technical details, and chaos. The Visual *Aide* in a technology presentation is a graphic, a chart, a pictorial, or any such construct that summarizes and compiles the *data* into understandable *concepts*. Whereas in a speech the visual *aid* complements the speech, in a technology presentation, the Visual *Aide* is the speech. It speaks for itself, and stays in the mind of the audience. It is a brain implant, implanting ideas, the wizard's ideas, in visual ways that words *cannot and do not* convey. You take the dense, hard to visualize data and create a cogent, visual picture for the audience, a visual image that can be grasped readily, understood, and remembered. That visual image must be a distillation of the essence of the data. It is neither summary nor compression. It is visualization.

In a technology presentation, there should be two or three of these surgical implants, these Visual *Aides*, that stay with the audience and teach them what they can never get from reams of data. There will be other visual aids, pictures, graphs and such, but only one or two Visual *Aides*.

However, each of the Visual *Aides* requires hours upon hours of creation, iteration, and preparation. They are not just word charts, or illustrative figures, or pictures of laboratory equipment, or graphs and charts. We discuss those separately. These Visual *Aides* are stand-alone stimulators, opening the mind of the decision maker to visualize your words. They may be projected with the latest technology, they may be animated, they may be still frame, they may be video. But regardless of the mechanism, more than anything else they create their own mental image that melds the mind of the decision maker to the mind of the wizard. Such a Visual *Aide* creates understanding. The right Visual *Aide* must supplant a thousand words, because no amount of words can convey the message.

An example of this Visual *Aide* concept, outside the technology world, is the political cartoon. The single political cartoon can be more poignant than an entire page of editorials. It gets an often complex, usually controversial, message across in a way that words cannot. It also fixates the memory. There is almost no data on it. It has been distilled to the "message."

In chemistry, there are symbols for the chemical bonds. There are single, double, and triple lines. There are hexagons and benzene rings. These symbols bring complex compounds and interacting forces into a common mental picture. A single line represents a complex type of bond called sigma. That sigma bond arises from shared electrons. Those electrons have different spin components, and so forth. All of that is shown in a single line.

Think of the double-helix DNA molecule. Can you explain that shape in words? No, you have to draw it out, color the adenine, thymine, guanine, and cytosine, and label it all. You have to show it from different angles. Francis Crick and James Watson received the Nobel Prize for their discovery of the structure. They explained that all the data resided on a double helix, so they drew a picture of where the components were and how they related. You can understand the concept without any knowledge of chemistry or biology. It is a Visual *Aide*. Once the structure was known, an entire field of modern science opened because, now, scientists had a picture they could connect with to make predictions and test against. It is a constant reference years after it was discovered.

In software, a flow chart depicts the logic of otherwise arcane code. The flow chart might be a type of Visual *Aide*. Another example is how Google goes about doing an internet search. All the explanations in the world would not be as cogent, as forceful, as a diagram of the logic. Now, these examples illustrate the power of Visual *Aides* to visualize concepts, thought processes, and interactions.

The right mental picture, the Visual *Aide*, not only takes the place of a thousand words, but it also distills your reams of data and your twenty-seven equations into something a decision maker can understand. The trick for the wizard, now, is to create these Visual *Aides* that link their technology to their decision maker.

A great example in physics are the Feynman Diagrams. Quantum mechanics is complex and difficult even to those who do it for a living. Richard Feynman created a Visualization *Aide*, the Feynman Diagrams, that make quantum interactions understandable, even to those who know very little about quantum mechanics.

Creating Visual *Aides* is not easy. It takes a long time, but it is the very best way—usually the only way—to distill data and make a persuasive argument to decision makers.

I was at a reception for Newt Gingrich, former Speaker of the House, and an outstanding communicator. Gingrich said three things that relate directly to this discussion.

Gingrich recalled the first time he met Ronald Reagan, the truly great communicator. Gingrich said they spent the entire hour discussing nothing but communication: how important it was, but how difficult, to put ideas together and present them in an understandable and convincing way. Getting the right message across to a decision maker is key.

The second element of interest involved a slogan Gingrich had just created. The slogan was, "They are the party of the food stamps. We are the party of the paycheck." The politics are not important to our thought, here, but his next statement is important. He said it took him "seven weeks" to come up with this idea and to formulate it into that one sentence. It takes time to distill concepts into cogent formulation.

The third item brings us directly to this idea of a Visual *Aide*. Gingrich told the story of his making a video with Lech Walesa, former President of the Republic of Poland. It was a video about Poland coming out of Marxist and Leninist oppression. Gingrich thought Reagan was the turning point, but Walesa said that was not so. Walesa said it was the nine-day visit, in 1979, of Pope John Paul II. That was what started Poland on the road to freedom. Pope John Paul II visited his former homeland, then held by Russia. It was a land without hope. One-third of all Poles saw the Pope in person. Millions more saw him on television. The Pope, in a country-wide mass, asked for God to "heal this land," meaning Poland. Whatever the Pope meant, the people took it, Walesa said, as a charge to break away from Russian oppression. Now, here is the important part. All over Poland, Lech said, shopkeepers began to put up signs that said, "2+2 = 4." The Russians knew the signs meant something, but they did not know what, nor what to do about it. What the signs meant was that there is a right and a wrong. That 2+2 do not equal 5 and 2+2 do not equal 3. Two plus two equals four and only four. In other words, it was only right that Poland should be free. It was totally wrong to be oppressed by the Communists.

Walesa said that this sign, this hidden proclamation of freedom, this Visual *Aide* ignited Poland. In 1989, Poland broke away from Russia and became, officially, the "Third Polish Republic." That was Lech Walesa's story. These Visual *Aides*, they are powerful stuff.

The point is that no one will read a viewgraph full of text, but the visualization says everything. That is what we mean by a Visual *Aide*. It takes effort to put data, equations, and formulations into an easily visualized concept, but it must be done to be effective. Practice doing this in every presentation.

A specific example of creating these Visual *Aides* in a technology presentation may help. I put together a presentation on intellectual property rights: patents, copyrights, software code, what we had to protect, why we needed to protect it, time elements involved, and the whole mechanism. Some of the steps in this process were business, some legal, and some technological. Some were easy for us and some were difficult. Some took a long time and others not so long. I produced a viewgraph of runners jumping hurdles at a track meet. The runners were different types of people (black, white, tall, short, fat, skinny, freckled, long hair, short hair, curly hair, etc.) indicating the variety of our products. Some of the runners were fast as lightning—the products I felt we should accelerate—and some were slow as turtles—the products I felt we should put on the back burner. (I needed to convince them of this, of course.)

The hurdles were labeled as steps in the intellectual-rights process. Some hurdles were low and easy to jump, and some were high, indicating that those intellectual-property steps were much more difficult for us. A few of the process steps required a pole vault, indicating that those were the ones where we needed funding (the *Why* of that particular meeting). Some had water hurdles, that is, penalties if one failed to make the hurdle. Some of the runners fell flat, showing failures in the past and where they had failed. The spacing between the hurdles indicated a relative timeline. The starting point showed where we were. The finish line showed where we needed to be. The decision makers were the referee holding the starting gun. I was asking them to "fire the starting gun." In other words, I was trying to convince them to agree to the process, release the funding, and give me the go-ahead to be in charge.

In this one Visual *Aide*, they had the entire presentation. As you can see, the words were difficult to understand, but the graphic had the impact of a well-though-out political cartoon—it grabbed them with the message. I used it throughout my presentation to reference to where I was in the discussion. I talked about "not sweating the low hurdles." I talked about needing "funding to make the pole vault hurdles." I talked about timelines. It was all on the viewgraph, my Visual *Aide*. I am certain that not a single individual at that meeting would be able today to recall any word I said. But I am equally certain that all would remember the Visual *Aide* if they saw it again.

Did this pictorial construct have business information? Yes, at a conceptual level. Did it have technology information? Yes, at a conceptual level. Did it forcefully convey *Why* we were meeting and what I wanted from the decision makers? Yes. Did it serve as a map and guidepost for the entire presentation? Yes. Did it stick in their minds when they were making their decision? Yes. Was it an indispensable Visual *Aide* to me? Yes.

But, as simple as this one example sounds, it took days and weeks to come up with the idea and formulate it for the meeting. However, it supplanted reams of data and made a sound, convincing argument, even though there was nothing truly technical in that entire viewgraph.

THE TITLES ARE KEYS

Let us continue with the idea of how to be selective and how to transform data into thoughts. Most presentations are given using projected viewgraphs. These are also called 'slides' because historically they were printed on transparencies (slides) and projected with an overhead projector. Hence, the pages of a PowerPoint presentation, or similar software, often are called slides. We discuss the software, later, but at this point just consider that you are creating viewgraphs by some method.

Each page, or viewgraph (slide), of your presentation should have a clear meaningful title, conveying a message. In a short presentation, each viewgraph might have a unique title, without continuations. If continuing pages are needed for the same title, then the title on each subsequent page should remain the same, except add "Cont'd" so that the audience can readily recognize that the train of thought is continuing. Change the title either when the train of thought changes or when you want the audience to think about the same thought in a different way. When continuing the same title, do not do this for more than two "Cont'd" viewgraphs, at most. It gets too confusing to go on with continuations. Break up the material into smaller segments, if necessary, and use different, but cogent, titles.

Think carefully about the titles. They are traffic cops directing the reader to the central theme of that viewgraph. A poor title gets the audience off track. A strong title focuses them. Carefully word the titles so that they precisely delineate the flow of reasoning. When you are delivering the presentation itself, accurate, well-thought-out titles serve as a tag to jog your memory to what is on the viewgraph and to know what is coming up on the *next* viewgraph.

I recommend that the title be a phrase, not a sentence. It can be two lines, if desired, but certainly no longer. One line is preferred. The meaning of the title, or phrase, should be understandable as stand-alone. When we showed the *Spider Diagram*, that particular page, or viewgraph, had a concise title that gave the message. It read, "Example Spider Diagram for Organizing Presentations." In your draft, pull out all the titles, separately, and read them, sequentially. They should flow smoothly from premise to conclusion along the strategic path you have chosen, the path you derived in Chapter Six.

You must practice the presentation so as to know what is next in the sequence. That way, you do not have to look continually at notes. Or worse, have to project the viewgraph, look at it, and then decide what it says. As you point to the projected viewgraphs, the titles give you the cue as to the topic being introduced. Keep the strategy in your head and avoid phrases like, "Flip to the next viewgraph and let's see what's next." "What's next" should be in your memory. If you cannot do this, make the last line of every viewgraph a jogger to cue you as to what is coming up on the next viewgraph, but do not make this obvious. If you know your strategy and have practiced, this should not be difficult.

UNDERSTANDING DENSITY

Density is how much mass an object has per unit of volume, how much stuff there is in a given amount of space, how much information you can put on one viewgraph. That is the *density*. How fast you present it is the *flux*.

Assume that you will be making a presentation to a group of about twenty individuals, a few of them decision makers of some sort. You expect to influence them to make a decision within a week or two. How much information should be put on each chart?

One of the problems with presentations is that often they supplant what should require a full written report. However, speed of commerce in most businesses precludes this report from ever being written and, even if it were, assures it would never be read by the decision makers anyway. Hence, the decision makers likely will base decisions almost entirely on the presentation itself. Any follow-up report will be filed immediately or given to subordinates who will read it without action. In addition to the speed of commerce, this generation "hears" with their eyes—flashy graphics, color, multiple monitors, large screens—vision and interaction supplanting detailed report reading. That gives the presentation all the more emphasis.

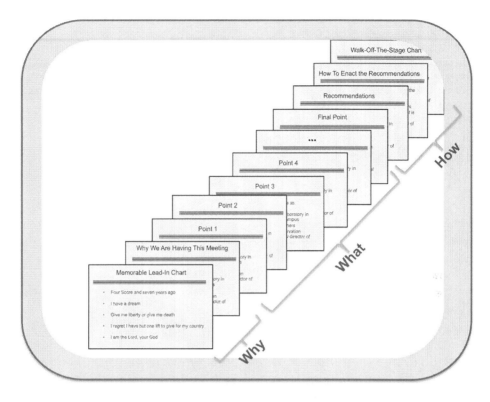

Figure 8-2: Titles Are Keys. When read by themselves, the titles of the viewgraphs should tell the complete story.

The presentation itself may become the only report and any documented reference is made back to the hardcopy of the presentation and any notes taken by the audience members. This puts a greater emphasis on how the presentation is constructed, organized, and presented.

Back, then, to density. How dense should it be? The traditional advice from marketing and social presenters is, for a normal conference room, (*not* a large auditorium) to use a font size of 18-24 point, have no more than three or four bullets (major points) on each viewgraph, and have no more than three to four words for each major bullet, with no sub-bullets. They are called bullets because a large dot, a bullet, often is used to delineate where the next idea starts. A sub-bullet is a bullet indented under a bullet to show a complementary, but subordinate, thought.

If you follow these guidelines, each major point will not be meaningful in and of itself. Each bullet will be a phrase to cue the presenter, but the phrase will not be understandable by the reader without the additional information given by the presenter himself. Afterwards, those bullets do not serve as notes for the meeting because those four or five words per line are meaningful only to the presenter and will be forgotten by the audience, immediately. Nor, as a whole, do such viewgraphs constitute a stand-alone document of the strategy and logic, because no one in the audience will be able to remember the explanation given with the words. What is the purpose of those bullets? They serve as surrogate notes for the presenter, not the audience. Rather than to elucidate the point, the bullets allude to the point by some phrase whose significance is known only to the presenter. If the decision maker has any hope or desire of making sense of the presentation later, the decision maker must take notes. Why? Because little of the information, none of the substantiation, and none of the details are on the charts. The details, the substantiation, the information are in the spoken words only. It is a rare decision maker who will transcribe anything more than the scarcest of notes.

This "standard" recommendation for viewgraph density has three major drawbacks. First, it requires that the audience listen with their ears, not read with their eyes, foisting a reliance upon the less trustworthy of the two senses. Second, it requires decision makers to take notes with some continuance and consistency so that they can refer back to them. I agree this is desirable and would be wonderful, if it occurred readily, but it is a hit-and-miss proposition. Plus, what they write is what they hear, which may not be what you said. A self-defeating thing also occurs; the higher up in the organization the individual is, the less likely she is to take notes, yet the more dependent you are upon her to make a decision in your favor. Third, when the decision maker or her team (who may not have been at the meeting) review this presentation, the intended thoughts are vague because of the paucity of original material and weakness of note taking. So, do not follow the advice of the marketers and social presenters if you are making a technology presentation to decision makers..

At the other extreme, we have those wizards who simply must put everything— literally everything—on the charts, fearful of omitting even the most insignificant of points. Keep in mind that the wizard, himself, does not believe the points are insignificant. He thinks every word, every detail, is essential. They are not.

I worked with a mathematician. Every bullet on his charts was a paragraph, every item in the bullet a full sentence, and every thought in the sentence at least a twelve-word phrase. The beauty of this approach was that it made wonderful reference material and could be understood even if reviewed years later by someone who was never at the meeting. But it was more like a final report than a presentation. The ugliness, and it is ugly, is the tedium and boredom it drills into the audience. It is tiresome to read all that detail during the actual presentation and most often that is just what the wizard does do. The wizard reads every word aloud to an audience that is cringing. The decision makers, reading their own copy and not listening, are pages ahead in *their* reading, removing their attention, and forming their own premature decisions. They become ever so fidgety to get on with what is important, not what they consider to be technology trivia.

The best approach for an audience of decision makers is somewhere between these two extremes. One should be able to read each bullet and understand its meaning, stand-alone, even someone who was not actually in the meeting. Each bullet should be a phrase, though, and not a sentence. Do not put periods in the presentation, even if it is a sentence as this adds busyness without adding meaning. The bullet itself sets off the phrases, or sentences, if they so be. The only exception to this is in the rare occasion when you want to use a bullet to make two statements. Usually, this is done by having sub-bullets, but a single sub-bullet may look untidy, so sometimes it is best to combine them if the second thought follows readily. The dash is another approach.

It is permissible to have two sub-levels, that is, a major bullet and two sub-levels of bullets. Each level, or sub-level, might have several bullets. As far as the bullet itself, sometimes a solid dot is used, then an open dot, a cross, an asterisk, a star, or other symbol to differentiate the sub-levels. Be careful in using a different style for the sub-bullets. More often than not, the different symbols add only busyness and not information. I recommend the same style bullet throughout the presentation with the next lower font size for sub-bullets. The sub-bullet should be indented.

When you get the presentation in final order, have someone read it, someone who has never heard you give the presentation. Let that person tell you how much is understandable and correct accordingly.

Contrary to what is usually taught, there should be very little "white space," that is, space without graphics or text, on the viewgraphs. Again, this is contrary to the school of thought that teaches that there should be no more than 3-4 bullets per page, each one no more than 3-4 words. This is incorrect for decision makers as it leaves the impression that one is presenting to a kindergarten class that reads slowly from a Big Chief tablet.

In general, decision makers prefer a more dense chart over a sparse one, but dense with high-level, interlocking information and graphics, not a dense collection of details. Keep in mind the central point of each viewgraph and make the material fit that point. Be concise, which is more than just being brief.

CHARTING A COURSE

In a presentation on Roman history or English literature, the audience begins listening from the very first word. The speaker may not show a single visual aid in an entire hour's worth of presentation. In contrast, technology presentations begin only when the first viewgraph is shown. All technology presentations have charts, slides, or viewgraphs of some ilk. This is not necessarily good, as it tends to make the audience think the spoken words are unessential. But it is the way it is.

I took training with a professional communications expert. She worked with a host of clients in entertainment, communications, marketing, and sales. Only a select few of her clients were technologists. She remarked how "frustrated" she was with her engineer clients, saying, "They rely on viewgraphs. Viewgraphs. And cannot communicate without them." Given her background, I understand the frustration, but she knew not a whit of technology. Her feelings only underscore and emphasize the need for a book such as this one. She had no knowledge and no appreciation for the complexity of the material and the difficulty of making a succinct, comprehensible, and believable technology presentation to a non-technical audience.

In a technology presentation, the audience will hear little before the first viewgraph is shown. That leaves limited time for verbalizing the basic premise and the *Why* of the meeting. The audience does not come together until the first viewgraph is shown. Thus, even the *Why* portion of the presentation must have viewgraphs. It is expected and part of the rules of play.

With the advent of modern graphics, small projectors, and flat-panel displays, it behooves the presenter to make the viewgraphs clear and understandable with text, graphics, and color. This can be overdone to a fault, so avoid cutesy and useless text animations, but do put major thoughts into written words and figures.

In 1987, Robert Gaskins realized the potential of the upcoming graphics processors and collaborated with Dennis Austin to create PowerPoint for the Macintosh computer. Later that year, Microsoft made its first acquisition by buying the PowerPoint company. Three years later, Microsoft came out with a Windows version and PowerPoint has since been the standard for presentations.

I agree with the ready criticism that a slick PowerPoint presentation often is a storefront without goods. Yale graphics expert Edward Tufte once remarked that PowerPoint, "elevates format over content, betraying an attitude of commercialism that turns everything into a sales pitch."

Mr. Gaskins, discussing the twentieth anniversary of his groundbreaking software, himself agreed, indicating that his original intent was that PowerPoint would be a summary to be followed by a full document. He called it an abomination that students often turn in book reports using PowerPoint rather than by writing. Gaskins and Austin enjoy telling the joke that the best way to paralyze an opposition army is to ship it PowerPoint, and thus contaminate its decision making.

One must agree with Mr. Austin when he said, "It's just like the printing press. It enabled all sorts of garbage to be printed." And as Mr. Gaskins worded it: "If they do an inadequate job with PowerPoint, they would do just as bad [sic] using something else."

Nevertheless, PowerPoint remains the presentation tool of preference. Like it or not, the PowerPoint presentation is rarely followed by a detailed report because the decision makers do not read detailed reports and the "efficiency" of the modern workplace leaves no time for full documentation and reporting. I find this regrettable, but see no significant changes on the horizon. I do not plan to war for an alternative.

Having said all that, PowerPoint and similar software tools are extremely powerful. Now, any wizard can have a professional-looking presentation in only a short time. Every technology presenter should become personally skilled in the use of PowerPoint and other similar tools.

For years, I have been privileged to work with companies that have full-time professional graphics departments. I work with the best and sing praises to their outstanding ability, professionals all. But, these people are not mind readers. You give them a concept, but it takes several iterations to get it "right" (in other words, to read your mind). You are far better off to develop your own personal PowerPoint (or similar program) skills and create the draft viewgraphs and graphics yourself. This will save you (and the graphics department) hours of work. They can put your draft in final form and resolve any remaining kinks in far less time than they can create from just your verbalized notions. If you do not have such a department, all the more reason to spend the time to develop your own PowerPoint skills.

For a wizard scientist, engineer, teacher, technician, medical expert or whatever, the real genius is in *creating* the communicative elements of the presentation—graphics that are compelling, pictorials that mind-meld with the audience, text that activates. Creativity is hard work. Many graphics and pictorials require professional software beyond the capability of PowerPoint. Again, I recommend putting all the draft material together, yourself, and using the professional technical publications department for the impossible pictorials and for the final polishing.

THE EFFECTIVE USE OF TEXT

How does one create persuasive text? The secret to writing is rewriting. Similarly, the secret to getting good text on viewgraphs is rewording, rewording, and rewording. Continue to work it until it is valid, succinct, and effective. If possible, make the wording of key elements memorable by pithy wording. Do not create a presentation of worn-out adjectives, examples being, "state of the art," "flexible," "readily upgradeable," "user-friendly," "up-to-date," and similar. After you compose the presentation, let the text sit a day or two. Then, review it. I am always surprised at how vague, ambiguous, and weak the wording can be when I read on Thursday what I wrote on Tuesday.

There needs to be consistency throughout the presentation. Be consistent with font sizes, bullet types, color and the like. A good text editor can be invaluable. Always use spelling check, always. Check, also, for those grammatical agreements. You do not want poor grammar to come before decision makers. It taints the good technical content and puts doubt in the mind of the decision makers regarding both your education and your experience.

THE EFFECTS OF COLOR

Color is a key element. I recommend every viewgraph be in color. But use color *sparingly* and only where it is effective. Do not produce "rainbow" viewgraphs. There is no reason to use different colors in the text font. It looks amateurish to have one line black, another red, and another blue, like the colored pencils used for coloring maps in grade school. This gives the presentation a circus look and is difficult to comprehend.

Certain colors change hues when they are projected on a screen and many of the pastels wash out. The color rendering by the projector is not a one-to-one mapping of the computer display—check this ahead of time to get what you want. You will not have time to do this in the conference room so use a laptop projector and see what it looks like. For the text itself, use dark blue or black as lighter colors can be ever so faint when projected in certain environments. Save the pastel colors for graphics, charts, and highlights.

The color red means "danger" or "alert," hence, its use in stop lights, emergency signals, flares, and muletas (that thing the matador holds in front of the raging bull). Red usually indicates an emergency or something wrong. Likewise, in your presentation, the use of the color red subconsciously brings up a panic or stress signal. Use a pure red color only if the emotion you want to evoke is, indeed, panic or stress. Use it sparingly. Here, I speak of the bright red, not the deeper tones of crimson or purple which *may* be acceptable.

Be careful of pastel and light colors They look fantastic on the computer screen, but even if bright enough, can be a different color when projected and thus give an odd effect. Practice your presentation with a projector and use the colors that are most effective when projected. Best to stick with the standard palette. Be especially careful of overlaying graphics on photographs. These can be disaster to read in a large setting where the projector brightness is marginal.

Some companies have gone so far as to define a standard color palette, standard fonts, and other such standards for their presentations. Check to see if the company has a standards-set and then follow it. The stereotypical wizard usually comes into the meeting not having checked these things, so mitigate the stereotype and use the standard.

Avoid shading and shadowing of the text. Such tools usually are perceived as gimmicky. They make it more difficult to read the text. Maybe for the first title page or short segments they are useful, but not in quantity and not with smaller font sizes (usually less than 16-point). Stick to traditional text and graphics and put your efforts into the content, not the form.

In every draft of the presentation, always number the pages. As you lay the draft on the table for review, shuffling to get the just the right order, you will appreciate those page numbers, especially when you drop the entire bundle on your way to the print shop. Hopefully, your presentation is not so homogeneous (read: boring) as to have every page look the same, but the page numbers do help for questions and reference. During the presentation, it is much easier for the big boss to refer back to page twelve, for example, than for her to tell you the title of the viewgraph she is asking about and you try to fumble and find it.

We are now into the "nuts and bolts" of making a technology presentation. You might ask, "Are these details really necessary in order to make a persuasive presentation? After all, making a successful presentation is not really 'rocket science.'"

Speaking of presentations and rocket science, there may be no greater name in rockets than Wernher von Braun. In WWII, von Braun, a native German, was a member of the Nazi party and an officer of the SS. During the war, he designed the deadly V-2 combat rockets for Nazi Germany. After the war, he and his team were brought to the US where he worked with the US military and later NASA. He became a US citizen whose crowning achievement was to architect the Saturn V rocket that launched men to the moon.

I was a senior in high school when I heard that von Braun would be speaking at a local university. He was world famous by then, having achieved his successes. I managed to get away from school and attend his lecture. He spoke about all the complexities involved, not just in launching a rocket into space, but in then being able to bring it back home and recover the crew without mishap. At that time, the capsule would be brought back to earth and would parachute into the ocean. The pick-up ship had only a few minutes to find the capsule and get the crew out. This was in the days before GPS, so there were military spotters scattered about looking to see where the capsule would land.

During the question-and-answer period, one of the university students asked, "When the capsule reenters the atmosphere, how close do you attempt to get to the location of the pick-up ship?" He hesitated only an instant before answering, "We aim right for the middle of the [expletive deleted] deck!"

Fortunately, for the US Navy, NASA was never on target. What he meant was that his team made every effort to be exact and perfect. Every successful endeavor requires that same attention to detail and that same knowledge of the essentials. You cannot make a successful presentation without following all these guidelines. Whether your subject is rocket science or not, these guidelines may seem like unnecessary details, but they are strong components of successful persuasion.

COMPELLING GRAPHICS

We spoke earlier of the Visual *Aide*, those constructs that coalesce complex material and transform it into understandable and cogent communication. There are usually only one to three Visual *Aides* in a technology presentation. In addition to those clever Visual *Aides*, one must show the data and formulations in some manner. Every presentation is enhanced by graphics—pictures, drawings, video, exploded drawings, 3-D renderings, pie charts, graphs, tables, scatter plots, bar charts, flow charts, cartoons, and the like. These give a rest from the constant cataracts of text. Graphics and pictures also transform the words into concepts. If the data are complex, always provide a graph or picture rendering. Use pie charts, bar graphs, photographs, drawings, and three-dimensional construction whenever possible. It is not always necessary to put the *actual* data onto the graphics, as sometimes the *conceptual representation*, such as an animation or drawing, is preferable. It depends on the audience and the material.

Videos, sequences of images, animations, and other mechanisms that present the data in different ways are good. If a video is interspersed in the presentation, make it flow smoothly without interruption. Do not permit 2-10 seconds of lag wherein everyone is waiting for the video segment to queue. Your audience of decision makers is quick to find other things to chase mentally. Do not give them an unintended intermission as a time lapse interrupts their train of thought. Make the video queue immediately. Avoid videos with lead-in music and swirling images that only serve as a queuing mechanism. They detract and subtract from your presentation. Do not fall prey to the changing whims of graphical arts designers. Your presentation itself is the lead-in, not the music and swirling videos. Queue the video *a priori* to mesh into your presentation like a gear. Most videos do not flow with the presentation because this is difficult to accomplish. Videos often have the effect of an interlude, which is not good.

The *PowerPoint-Commando* is a denigrating appellation assigned to presenters who have more flair than content. They mistake fancy fireworks, glitz, and glitter for technical content. If the presentation is technically unsound, clever videos, animation, and flashing lights are no more useful than putting makeup on a pig. *PowerPoint-Commandoes* create elaborately ornamented text, dynamic viewgraphs, and flashing visual displays, but the underlying content is not compelling. To decision makers, content is everything. Weak material is weak no matter how eloquently expressed. Job one is to get the content right.

PowerPoint-Commandos have two favorite tools. I recommend neither. They are clip art and dynamic text. Most software presentation products, including PowerPoint, have a library of clip art. These are sketches, drawings, paintings, or images of a general nature that can be inserted readily into the presentation. Mostly these are just cutesy and fill the white space. If you really feel you must use them and cannot create your own, use them only as a last resort and only when the clip art uniquely adds *meaning* to the presentation. Do not settle for window dressing. Do not get into the habit of inserting some trivial clip art on every viewgraph as an attempt to make it "entertaining." Use only the clip art that really explains the data and use clip art only when you cannot provide your own graphics, those indigenous to the research or presentation.

The second tool, dynamic text, takes two general forms. The first is where the text "builds." In this use, the speaker presses the control and the next bullet appears, for example, so that the speaker is talking to and about the *current* highlighted bullet while the other previous bullets only weakly appear. In using this technique, the intent of the speaker is to keep the audience focused. This is a good thought, but it does not work. The audience is mostly distracted by the constantly changing screen—members may still be thinking about the prior bullet and it goes faint. If the presenter goes too slowly, the audience keeps wanting the speaker to hurry up as they already have the message. Because the audience is made up of diverse people, there is no correct speed, so the net effect is not positive. Avoid this "build" capability with text and simply show the entire page at once. You do not plan to spend all day on it, anyhow.

While it does not work with text, this capability to show information in layers does have considerable merit when explaining a complex *graphic*. By *layering* the graphic, you can lead the audience through the train of thought you desire them to take. When the graphic layers are built in a logical sequence, it can be very effective. But, use no more than two or three layers of "build" and use the layering capability only if it really aids in clarity.

Another form of dynamic text is to have the text rush in from the outside as it enters. It may sit there and wobble to bring attention to itself. Or it may leave with an explosion, vanish magically in a cloud of smoke, or zoom off like a race car. It may sit there and vibrate or wave its hand to get your attention. This is silly. Avoid all of these cheap gimmicks. They detract from a professional presentation. Better to spend your time creating concepts.

SPECIALIZED AND STANDARDIZED VIEWGRAPHS

The term Quad Chart, or four-square chart as it sometimes is called, is a viewgraph bisected with two orthogonal lines to produce four quadrants, usually of equal size. Different information is placed in each quadrant. For example, one quadrant might have an overview of the project, another the financial status, another technology status, and the fourth the schedule. This style is best used when there is a standard format such as with financial reviews or periodic evaluation, especially when much of the information remains the same from review to review and it is updated only infrequently

If the presenter wants to put a large amount of information on a viewgraph that might be used as reference, (i.e., not shown, but included in the hardcopy), the Quad Chart is a good technique. If you use this approach, have some pictorial representation in at least one of the quadrants and a graphic or schedule in another. A single page of dense text in four quadrants is too formidable.

Most organizations have standardized formats for financial data and schedules. Go to these departments and find out what is standard. Use the same format in your presentation. If possible, get that department to endorse your information. If not, still use the standardized format and do not create your own. While it may seem straightforward to you to expect that the audience make the transition from your unique creation to the standardized form, decision makers will not do this. There are at least four reasons why you must use the standardized formats. First, if you do not use them, it will show the decision makers that you made no attempt to get the right format. Second, it will show them that you spent unnecessary and unproductive technical time creating a new format. Third, it will reinforce their image of wizards being unknowing and uncaring of the real world. Fourth, decision makers will not, or cannot, make the transition to your format. They will dig in their heels and not accept it, no matter how simple you believe the transition to be. The conference room is not your world, it is theirs. You will spend precious time explaining something that could and should have been avoided by using the standardized formats.

Pie charts, bar charts, line graphs, and similar are all very useful. But, choose wisely. Develop some sense of knowing which type of chart works best for which type of data. Non-technical persons are accustomed to having the independent data on the abscissa (x-axis) and the dependent data on the ordinate (y-axis). They may not know what those words mean, but do not change the orientation of plotted data unless it is absolutely required. Time, for example, is traditionally on the abscissa. It will be confusing to decision makers to put it on the ordinate.

BACKUP MATERIAL

It is common practice to put substantiating and supplemental material in what is called "backup." These are usually viewgraphs placed behind the last viewgraph presented. The Backup viewgraphs are intended to answer or expand on potential questions. Or, they can be used to take the place of a follow-up report and simply give more detailed substantiation, which is more common. The Quad Chart, discussed above, is an example of this application. Backup viewgraphs are a carryover, I believe, from when a presenter used a Vu-Graph or similar hardcopy projection machine and the presentation material was on transparencies. Used with that media, the presenter could shuffle or insert these pages as required and it made the presentation appear seamless.

When this practice transitioned to the computer viewgraph show, it lost much of its effectiveness, because it is not possible to shuffle viewgraphs seamlessly on a single computer. If you change the predetermined sequence, *in situ*, the audience must accompany the presenter on a mouse-click-and-pick journey to find the right backup viewgraph. This maneuvering is distracting to the train of thought. Avoid this practice and go to your Backup only as a absolute last resort, after you have tried other techniques, discussed herein. Change the order of the presentation *only* if you feel it is *absolutely* necessary to maintain audience coherence. Remember, though, that once you do this, you have lost control of the audience and persuasion is more difficult. Ensure the digression is worth the sacrifice.

I have seen wizards go through the entire presentation, complete it successfully, and then say, "Let's now go through the backup viewgraphs to see if there's anything I've missed." What? Please, do not do this, ever. The *only* real use for the backup viewgraphs is for the audience to have them in their hardcopy as reference, and even then, *only* when necessary. Do not brief these backup viewgraphs if there is any reasonable way to avoid it. I recommend you avoid the problem entirely by having no backup viewgraphs whatsoever. If there is supplementary material you want them to have, package it separately and do not call it backup material.

ORDER AND SEQUENCE

Bouncing between disjointed viewgraphs is disruptive. The audience cannot follow the flow and it creates confusion. Page through the presentation ahead of time to ensure that the order is what you want and then stick to it.

If, in briefing the selected sequence, a member of the audience asks a question about a prior chart, what do you do? Most amateur technologists immediately page back to that chart and brief the entire chart again, even when the interrogator admonishes, "You don't need to go back to the chart, but I have a question." Most wizards, if they go back to the viewgraph, simply cannot resist the temptation to brief it again, often taking more time than they took originally. Valuable minutes are lost for nothing. Answer the question succinctly and do not page back to a viewgraph unless it really is required. If done properly, the audience has a hardcopy already so they can look back at their copy while you answer the question. (*You* should know what the chart says without going back, yourself.) By all means, answer the question, but go back to the chart only if the question really is important, or the person asking the question is a major figure, and even then only if the question cannot be answered adequately otherwise.

The order of the presentation should follow the *Why, What, How* paradigm we discussed and the strategy you developed in Chapter Six. The titles of the charts should ensure that the flow is logical and the transitions smooth, as we discussed in this chapter. If you change the order the night before, as I often do to tailor to an audience, do not forget to examine the titles and change them, as required, to get the flow right. If the order is changed, the titles may require modification.

A fast way to establish or validate the order is to print a hardcopy of all the potential viewgraphs and lay them out on a table. For long presentations, you may have to carpet the floor. This lets you stand back and see the big picture. You quickly will spot flaws and inconsistencies that were not identified when flipped through sequentially on the computer or projection screen. Shuffle the papers around on the floor until you achieve the order you want. Change the titles to make the flow right. Modify the changes in the computer. (You will be delighted you numbered them.)

The hardcopy material you distribute during the meeting, and plan to leave with them, must be identical to the presentation. Otherwise, confusion. Always number every page on the presentation and in the accompanying hardcopy. Make sure your hardcopy pagination matches the final order of your presentation. Hardcopy itself is discussed in Chapter Eleven.

Presentation programs such as PowerPoint have a tool that permits every viewgraph to carry a header and a footer. The header usually carries logos and company information or similar data that do not change and are usually not a problem. The footer, on the other hand, may contain the date the presentation was compiled, or a version number, or other unique, but changing, identifiers. From a presenter's viewpoint, this is a welcome aid to differentiate versions of the presentation, maybe a different sequence of viewgraphs, a different emphasis, or just different audiences. However, if you put this information in footers, be sure the dates are correct on the viewgraphs you show and on the hardcopy you deliver. Be sure you want the date included.

Having any date appear other than the current date (date of the presentation) leaves the impression of being old information that was not updated. I was given a presentation from a company seeking investment. The presenter was a vice president of that company. At the bottom of each viewgraph was a date two years prior. The date may just have carried over from other viewgraphs or a template and gone unnoticed. Or, maybe I was just getting old information? Maybe they had been trying to sell this same idea for two years with no takers? The date levied a heavy question that took me away from the presentation. I never asked about the date. It is not my job to make the presentation for them. It could have been a clerical error. It could have been a format error. It could have been old information that was dusted off to "fish" for interest. I did not, and still do not know, and that is the point.

Distilling the Concepts

- Learn to be selective.
- Know more than you need to know, but do not use it all.
- Keep within your domain knowledge.
- Know the power of the Visual Aide. Take the time to become proficient.
- Use color sparingly.
- Create effective titles that map you through the strategy.
- Understand what is meant by content density and optimize it.
- Use meaningful, cogent, and understandable graphics.
- Do not brief any backup material in your presentation.
- Be certain the sequence of the viewgraphs follows your strategy.
- Keep headers and footers current.

Chapter 9: Targeted Communications
Articulating Those Concepts, Algorithms, and Equations

A word fitly spoken is like apples of gold or pictures of silver.
— The Bible, book of Proverbs

William Bennett, educator, author, and former US Secretary of Education frequently remarks, "Words matter." And so they do. Select your words carefully. Every sentence need not be pithy, but each does need to be precise, concise, cogent, unambiguous, and understandable. Your articulation need not excel Shakespeare's, but words do matter.

With words only, poets and novelists create images in the minds of their readers. Open a good novel and within minutes your mind will quick-draw against desperadoes, meander through misty fjords, search for a lost love, or rocket onto a distant planet. You will vicariously become the author's protagonist.

There is vitality in words. The technology presentation must have clarity of words and the proper selection to make complex technology concepts crystal clear to decision makers. The decision makers may not walk in your steps, but they must travel to the same conclusions. Use the wrong word, or an ambiguous one, and you befuddle the audience and they stumble toward rejection. Ambiguity, a boon to politicians, is an enemy of the technology wizard.

There are and have been gifted wizards such as the late physicist, Richard Feynman. Feynman could spellbind lay audiences and make understandable even the most complex ideas of quantum physics. When Feynman spoke, the audience came away with a solid appreciation of the science, regardless how shallow their true depth of understanding. Why? First, he knew the subject very, very well. He helped research quantum mechanics. But this is only a part of the story. There are many, many people who know a subject well and yet are not able to communicate to anyone other than peers. What Feynman added was the genius ability to transform complex concepts into palpable images. He used words to transform the unknowable into the knowable. Words will enable you to transport a decision maker from ignorance to epiphany, from shadow to light. This is the task of the wizard presenter—to take complex technology concepts and transform them into relationships, comparisons, and metaphors understandable by the non-technical audience. Words matter. Choose them for their clarity of meaning and tailor them specifically to the audience.

DYNAMICS

I took golf lessons. I was given a club, a tee, and a ball and told the steps to go through. I memorized the steps. I teed the ball up and thought, "Head down, elbow straight, make the swing go in an arc, keep the knees slightly bent, right elbow against the side, fingers locked, stand with the ball slightly forward of center, follow through, keep the eyes on the ball." Maybe this is where you are now, thinking about that next presentation. You are reading the steps in this book and thinking through them.

If what you plan to do is read the book and make a presentation, you may end up like I did on the golf course. You swing with mighty energy and the ball remains untouched. You try again. The ball pops up and rolls a few inches to one side. Better. You discover that it will take practice on each of the steps. You practice with and without the ball. In other words, you can practice the steps without actually giving a presentation.

Arnold Steinhardt is first violinist of the Guarneri String Quartet. He relates of his lessons from the master instructor, Ivan Galamian. Galamian desired Steinhardt to develop certain complex finger motions of the bow by practicing for weeks *without the violin*. Similarly, you can practice elements of technology persuasion without actually giving a presentation.

After you practice golf, violin, or any other skilled task for hours and days and weeks, you find you no longer think about the details because they come naturally. At first, they hinder the game, but once developed they strengthen it. The same with technology persuasion. Learn the techniques. They may seem awkward and different at first, but the techniques and methodology are required. Once they become natural, you will find that you can implement your strategy successfully every time.

Let us begin with *how* you say the words. Do you mumble? Do you let the end of your sentences taper to a volume of zero decibels? Do you make presentations in monotone? It is not sufficient just to know *what* to say; knowing how you say it is equally important. The message will fall flat if the delivery is flat. The words must have articulation. The words must have life, emphasis, excitement, and dynamics. You must know how to say the words and put them together. This ability to speak, in and of itself, is so powerful that many salesmen have hugely successful careers based solely on *sounding* convincing. They know very little about their products and have very little product to sell.

I am not advocating becoming the stereotypical salesman, and it would not work with decision makers anyway. I am not advocating a hollow presentation devoid of information and facts. Nevertheless, it will not be convincing to most decision makers unless you can make it *sound* convincing. As a minimum, you must hold their attention long enough for them to think about what you are saying. And, thinking is hard work.

Maybe you do speak somewhat in a monotone or mumbling, but you think that is "just the way you are," that it is "okay." It is not. Period. To illustrate, try an analogy. Let us see how a monotone voice would appear if it were manifested in the other senses. Try vision. It would be devastating to suddenly find yourself blind, but what if you were not blind? What if, instead, you had perfect vision, but the only thing you were ever allowed to view was a single, unchanging image in front of your eyes. You can look around, but you always see the same thing. What good would vision be if it all looked the same? Does this give you an idea of how monotone articulation can come across to the audience?

Try the sense of touch. Suppose you have been blessed with a phenomenally sensitive sense of touch. The only problem is that everything feels alike—your spouse, a dirty dish rag, a garden slug, all alike.

Try the sense of smell. You have a great sense of smell. You can detect three parts per billion of any odor. The only problem is that everything smells like ammonia.

Finally, try the sense of taste. You have the most delicate tastebuds of any human on earth. You can taste something in the smallest of quantities. That is the good news. The bad news is that everything tastes like Spam. Every drink, every snack, always the same. Apple pie Spam.

Now, back to the ear, with its sense of hearing: that sameness is called monotone—same tone, same pitch, same volume, same everything. Or, it can be mumbling. Same, same, same. Oh, it drives you nuts.

Put some inflections and emphasis in that voice. Get excited about something! Inflections need not compete with a kindergarten teacher reading fairy tales, but inflections are needed. Practice the delivery until it becomes natural, but has some life in it. Change the pace, the rhythm of the words, sometimes faster, sometimes slower. Make the pace match your emphasis in the presentation. Err on the side of enthusiasm. Very few wizards can ever exaggerate to the point of sounding like drama queens.

Another common trait is trailing off in the volume at the end of the sentence. You start out fine at the beginning, but you keep going on and onuntilthereisnoendatall. Practice keeping the volume at a sustained, higher level. You choose words carefully, so you do not want the audience straining, mistaking them or not hearing them, at all.

ACCENTS AND ENUNCIATION

I took a course in silversmithing from Dorothy, a spunky, saucy, old lady. We students huddled over our torches and acid flux, awaiting our first real try at silver. She distributed to each a long, thin rod of sterling. "Make this slender rod into jewelry," as her only instruction. Around the tables, the eight of us looked at each other, fired up our torches, and started to work. We chopped our little silver spaghetti into different lengths, soldered crosses, bent circles, formed loops and hoops, and spirals and ellipses.

In awhile, Dorothy judged our work with her scowl. (She could be testy at times.) "You people are boring," she said. "Look at what you have made." All of us looked about. There were arcs, ovals, crosses, and snakes, all looking like little silver rod segments soldered into arcs, ovals, crosses, and snakes. "You're not thinking," she said. "You're one-dimensional. Your work is boring because *round* is boring."

She was right. Little round contorted rods look like little round rods contorted. We had made something out of little round rods because that was what she gave us. So, we understood then that there were other "dimensions." (Artists use that term without meaning spatial dimensions. A physicist would correctly call it "degrees of freedom.") Over the next week, we pounded those rods into chips, twisted them into DNA molecules, engraved designs, layered elevations, enameled cloisonné, forged them into maple leaves, and etched them into burnt French fries, anything but round. It was beautiful. We were artists. Our work went from craft to creation. This same transformation can occur in your presentation.

Nothing will put an audience to sleep faster than plain vanilla. Most individuals have a natural voice of interest, so use it to good effect. Articulate and emphasize. Intonate and add dynamics.

Use your natural speaking voice. With daily access to global audio and visual productions, there is reason to think that the best voice is one without dialect or regional flavor. But, how many gallons of vanilla can you stomach, hour after hour? Think of the most effective speakers you know, the most influential, and the ones that hold your attention. Bring to your ears your favorite vocalist. Whether your choice is rock, classical, country, bluegrass, or whatever, I would be willing to bet that your favorite has a distinctive voice. There is something different about it that draws you. It is not the sameness. Those individuals use distinctiveness to advantage. So should you.

I work with two people who have pronounced Virginia accents, one a physicist and one an attorney. Both have extremely fascinating delivery and both are highly successful. The attorney speaks almost too slowly. However, he never uses an "and" or a "but" and his pauses only give you time to think. He is captivating.

I know a great software engineer plucked from the Bronx. His delivery is so rapid I have to hang onto my Texas seat to follow it. It does not distract from his content, it adds flavor.

Admittedly, there is stereotypical association that comes with a Boston brogue, a British enunciation, and a Texas drawl, as examples. But, by and large, these accents can be effective in making the presentation colorful, dynamic, and interesting to the audience. Some years ago, I queried a professional speech coach regarding the implication of accents, whether certain individuals should consider phonetic training to eliminate accents. Her advice was, "No." She was firmly convinced that accents gave the voice color and that a Hollywood "sameness" should be avoided. I have validated that advice. I find no correlation between success and language accent, as long as it is clearly understood. Use your best voice, your natural voice, to advantage.

That is not to say that poor grammar, inadequate diction, or local slang is acceptable. By no means. The quickest way to turn off an intellectual audience is to use incorrect grammar. Get the grammar right. Work at it. You can have all the Texas accent you want, all the Boston brogue there is, but eliminate expressions such as "I'm fixin' to," or, "I'm gonna," no matter how common they may be in Sweetwater. "Go figure." "Me and Jim, we..."

Many US companies, especially technology companies, have a large workforce of varying ethnic origins. Accents are expected. But the words, with or without accent or dialect, must be understood without hesitation by the audience. A few seconds lost in translation can deter the message the speaker wants to deliver. Regardless of whether the accent is regional or foreign, ensure that the words are understandable to an audience unfamiliar with your accent. If the words are readily understandable, I maintain that the accent and local flair are most effective. If the accent is so strong that it is not understood, then work with a coach until the accent is characteristic, but not distracting.

I was in Bangalore, India with a native scientist I had hired only six months prior. Dr. Ramon spoke English, but he was difficult to understand when he spoke rapidly. At lunch that day in the heart of Bangalore, one of Dr. Ramon's business colleagues joined us. The associate wanted to discuss outsourcing, said Dr. Ramon. After introductions, we sat and began a very nice lunch. Dr. Ramon and his counterpart began to speak to each other in what I assume was Hindi. I have no knowledge of Hindi and after about five minutes, I rapidly lost the ability to look interested. I wondered that they continued their protracted dialogue in Hindi and why these gentlemen were not discussing potential business relationships in English. Indians are extremely sociable and polite, so this was very much out of line with what I had been accustomed to there. While I was spooning my soup, I caught one English word. I kept listening. A few moments later, another. Listening carefully, I heard another. An epiphany occurred. They were not speaking Hindi at all; they were speaking *English*. But, because their accents and pronunciations were so abstruse, I had understood only a few words. Such unordinary articulation was totally unexpected. I strained both eardrums to understand what they were saying, but finally settled for nods and smiles of feigned acknowledgement.

That evening, I politely inquired of Dr. Ramon why he could understand the accent and I had such difficulty. (I did not tell him I understood next to nothing.) His reply was that many students from poor Indian regions do not have satellite broadcasts and cannot attend schools taught by native English speakers, so they learn from each other. I take no responsibility for the truth of that statement but, I can assure you, I understood only about ten words in as many minutes. In an analysis done by the *Wall Street Journal*, they attributed the cause to the entire education system, not only in India, but in similar outsourcing countries. It appeared to me that a so-called English dialect had grown up as the people could understand each other yet not be understood by native English speakers.

This is not an accent problem, it is clear mispronunciation and to such an extent as to be unacceptable in a presentation. Such mispronunciations must be corrected to make effective presentations. In some respects, this is a very tolerant time period in which we live, but as a technologist, you require effective communication, not tolerance.

This anecdote in language compares to the quantum mechanics graduate course I took years ago. As if the subject matter were not difficult enough, the professor was in the US with minimal English skills. His knowledge of quantum mechanics was excellent, I suppose, and I am sure he eventually developed the needed English, but I had his course the first year he came into the country. His inability to articulate made that course an unbearable, unnecessary, and unforgivable obstacle for the students.

The point of all this is if you are easily understood, then you have what it takes to be a good speaker. If you are not, then work diligently on that until it is, at least, acceptable. Do not mistake tolerance for persuasion. Take your natural voice, but use it effectively.

SPECIFIC WORDS OF CAUTION

Two words you must eliminate, at all costs, are "and" and "uh," particularly in juxtaposition. This annoyance sometimes occurs even with speakers who should know better. Practice the presentation and listen to yourself. When you feel the urge to say "and, uh," just stop and say nothing. Silence is golden. It is effective. Learn to pause. The pause does not have to be pregnant, just a pause. That pause is far preferable to saying something meaningless like, "and, uh." As a speaker, the pause seems overly long and noticeable, but it is not. If the audience is listening, they will appreciate a moment of reflection to absorb the fine words you have spoken already. If they are not listening, the momentary pause may bring them back to consciousness. Either way, it is a positive effect. When you practice, have someone give you hand signals if you say "and" or "uh." It will be distracting at first, but this will help you focus. Substituting the pause will soon remove the superfluous phrase. The pause will remain and will introduce natural breaks in thought, which is a good thing.

Ivy League debaters use the occasional stuttered "and, uh" as an effective part of their debate response—to feign defense while thrusting the touché. Technology wizards are best off eliminating the "and, uh" entirely.

In a culture ever more sensitive to diversity, there are many people who are truly offended by vulgar and profane language. Eliminate vulgar expressions and profanity. Strong profanity is a bullying of rhetoric instead of communicating with precise ideas. Vulgar and crude remarks may bring a laugh or chuckle, but the internal reaction usually is not humor. Stick to describing the technology in the language of the professional.

Eliminate words or phrases that reflect sex, race, religion, or stereotypes. There is no joke so funny that it escapes this admonition. It is a nation of social and political hypersensitivities. If you make presentations outside the US, this is even more important. If a key decision maker is offended, it will be almost impossible to get support. A technology presentation should be just that, a *technology* presentation.

The use of "he" or "she" is problematic. *"He/she"* sounds awkward if you say it that way and meaningless if you write it often. It becomes silly. And yet, to continually use "he" is also objectionable. The use of the word "they" when a grammatical agreement on number can be smoothly substituted is a solution, but often this cannot be done without sounding stilted. Using "they" when it does not agree in number is just poor grammar. Until a better solution comes into accepted English usage, I suggest varying the terms, one reference using "he," another reference using "she." Or using the neuter-free word "one" as in "one should interpret the results in this manner."

Gender references are imbedded in older terms such as "chairman" and "policeman." Use the modern versions of "chairperson" or, simply, "chair." "Police officer" is a non-offending substitution. This is a sin of both men and women. It is best to remove the gender-specific words from your vocabulary altogether.

ACRONYMS AND JARGON

Avoid acronyms altogether unless you are certain that *everyone*, absolutely everyone, in the audience, will understand quickly what they mean. The words "everyone" and "quickly" are the operative words. Most technology areas are replete with their own unique dictionaries of acronyms. Do not assume that these acronyms are widely known. Even if they are, sometimes the audience will hesitate in their minds to come up with what the acronym means and you will lose their focus. Obviously, acronyms like CEO and CFO are known by decision makers and used frequently. But do not try CPSK, SCMA, PHIGS, PQFP, SACSA, or DADA. Even if you define them, they can get forgotten quickly. If you are in the hospital, words like ER and PDA might be common. Know your audience. In any environment, use acronyms only if you absolutely must, and then sparingly.

So, when is an acronym acceptable? If a lengthy phrase is required and there is a common acronym for that phrase, then say the broader phrase in the beginning and introduce the acronym. The same applies for a phrase you must use often. As you repeatedly use the acronym, ensure that you also repeat the broader phrase at least two or three more times. This will reinforce to the audience the meaning of the acronym. You can see how awkward this becomes if you introduce many acronyms, so do not introduce them.

If the acronym is something like TCP/IP, it is probably better just to say it is a set of protocols designed to get data from one device to another when connected over a network. That is, describe what it does, in general, rather than say "transmission control protocol, internet protocol," as if that would help. The bottom line is this. Do not assume that just because you define it, an audience of decision makers will know what it is with any degree of familiarity. Just because it is commonplace for technologists, do not assume that decision makers will have any real sense as to its meaning.

Avoid buzz words and jargon. They really do sound nerdy and create a barrier to communication. Such terms label you as stereotypical. I am convinced that many wizards persistently use these terms to try to appear superior to the decision makers. Do not do this, whatever the reason.

COMMON PITFALLS

Did you play a team sport like football, soccer, basketball, or volleyball? The team has an overall game plan, usually based on a film and scouting analysis of the particular opponent. The team develops plays. Within each of these plays, each player has his own specific part, linemen here, cornerbacks there, center, guards, whatever. Perfect. Let us run out and play some football and beat the other team. Whoa, whoa, whoa. The game plan and plays are meaningless if we have not developed some specific skills to go with them, skills such as tackling, blocking, catching the ball, handing off the ball, knowing where the down markers are, knowing how to defend, and so forth. Techniques take effort to master, but no master game plan can be effective unless each player knows the required techniques. The guidelines we bring up are like those techniques. Work with each to master them.

Avoid meaningless words or words that are made meaningless by repetition. After a long day of listening to mostly sales presentations, I looked around the room and noticed that three members of my staff had lost interest in the presentation and were playing a game they made up. They scored points whenever they heard a speaker use the words *user-friendly, viable, one-of-a-kind, innovative, value-driven, latest, modular, easily upgradeable,* and so on. I resisted the urge to play, but with difficulty. Words become meaningless from overuse and misuse.

The word *data* always occurs often in technical presentations. *Data* is plural, *datum* being singular. But it sounds ever so awkward to say "the data are" this or that. Since *data* can be singular or plural in construction, I recommend using the expression that calls the least attention to itself since the meaning is never misunderstood. In formal writing, however, stick to the rules.

I do not know why technologists and scientists have such a propensity to use the non-word *irregardless*. *Regardless* already means *without regard*. Try the word *irrespective* if you simply must have a prefix.

In a similar vein, the word *unique* does not permit comparison. It cannot be almost unique or most unique or nearly unique or "uniquely one-of-a-kind." If it is unique, it is already one of a kind, the only one of its kind, having no like or equal. It cannot have qualifiers.

Do quantify statements and assertions to make them definite and exact. "We increased productivity by 53% this month." Or, "The data show a 20% improvement in noise reduction." Avoid indefinite and vague words such as *quite, a lot, many, few, most, practically, very, sometimes, often, probably, plenty of, numerous, several,* and the like. It is, after all, a technology presentation, so quantify and tell exactly how many or how long. No need to be a Mr. Spock to the fourth decimal, but quantify to the level that is appropriate.

Avoid the superlative unless you can and do quantify it. Saying our product is "the best" may seem defensible, but in what respects and how? "Our product is three times faster and half the cost of our closest competitor" is far preferable. Of course, you will be expected to be able to back up such assertions.

Avoid saying you *proved* such and such, like "we *proved* that the public preferred our product." *Proof* has a very special meaning in mathematics and science and unless you are describing such, do not use the word. Scientists and technologists do not use the word "prove" unless they use it in this way. Show the data and speak to the conclusion, but do not say you *proved* something.

Turn negative statements into positive statements and conclusions. It is easy to complain. It is easy to distribute blame. It is easy to elucidate shortcomings. It is also easy for the audience to take the counteroffensive. So, do not speak from the negative side of the equation. A "negative" version sounds like, "We do not have sufficient funds to purchase needed supplies." "We do not have enough people to complete the research on time." A positive version, which says the same thing and so much more is, "We can buy one hundred of the new physics texts for only $2,000 additional." "We will complete the project by May 10, if we add two additional radiologists." This habit of stating things in a positive, quantifiable way, invariably moves you from abstract to the concrete. Making positive statements is more difficult, requires more thought, and does not come naturally to most. Start by putting your presentation together as you normally would. Then, go through and methodically convert every negative statement into a positive one. This will focus your assertions and the favorable habit will develop.

Use spelling check.

If you do not know the difference between *your* and *you're*, between *its* and *it's*, or between *to* and *two*, have the material proofread by someone who does. However, you should learn the basics of grammar. *The Elements of Style*, by the venerable William Strunk and E.B. White, although decades in use, is still one of the most concise books for quick reference and learning. It is not slick, but it is short.

CHOOSE THE RIGHT WORDS

Words matter. Spend time developing a good, solid vocabulary. Not necessarily a knowledge of technical jargon or a knowledge of esoteric words, but develop a knowledge that can choose between "affect" and "effect," between "hypothesis" and "conjecture." Few things are more pleasing to the intellect than the choice of the right word. The master of words, the late William F. Buckley Jr., was accused in a debate forum of using a longer, more arcane word when a simpler word would have meant the same thing. The wry Buckley countered by alleging his five-syllable word maintained a rhythm that would have been lost with the shorter, three-syllable word. You do not need to achieve that level.

It is powerful to have the ability to "turn a phrase," so to speak. This ability to say meaningful things in memorable ways is a valuable asset to a speaker. Whether speaking or putting the presentation into text form, work ever so diligently to select the word or phrase that precisely connotes your meaning. Avoid being trite or cute.

In his *Confessions,* Saint Augustine describes a renowned and brilliant orator, Faustus, with whom Augustine has waited years to discourse. He is delighted to learn that the venerated Faustus will finally travel to Carthage and converse with Augustine and other scholars. The famed speaker arrives, the students are set, and Faustus begins his discourse. Augustine is bitterly disappointed, writing,

> "[The ideas Faustus presented] *seemed to me none the better for being better expressed, nor true simply because they were eloquently told. Neither did I think that a pleasant face and a gifted tongue were proof of a wise mind. A statement is not necessarily true because it is wrapped in fine language or false because it is awkwardly expressed."*

Note that Augustine talks to both sides of the issue. Expressing a concept with pith and wit does not make the concept true. Contrariwise, expressing a concept poorly, with poor grammar does not make the concept false, but neither does it make it understood. The former leans toward something that is labeled "spin," trying to convince an audience using élan, wit, or intimidation instead of content. This is the stereotypical "salesman" approach. This error is not often encountered in wizards, as they are prone to be brutally honest. The problem child is the latter error, the inability to express true, but complex, technology issues in any way that decision makers can understand.

Earlier, we talked about the great, moving speech by Patrick Henry at the Second Virginia Convention and his final, stirring phrase we all can recite, "Give me liberty or give me death." It was a moving exhortation and an example of choosing the right words. However, in general, as a speaker, Henry had a fatal flaw. Here is how Thomas Jefferson described his speeches. "His eloquence was peculiar, if indeed it should be called eloquence; for it was impressive and sublime, beyond what can be imagined. Although it was difficult when he had spoken to tell what he had said, yet, while he was speaking, it always seemed directly to the point. When he had spoken in opposition to my opinion, had produced a great effect, and I myself had been highly delighted and moved, I have asked myself when he ceased: 'What the devil has he said?' I could never answer the inquiry."

I have heard technologists who were in the same club. They were awe-inspiring when speaking and everyone loved to advertise them as keynote speakers at a conference. Yet, afterwards, one remembered only the impression, not the movement, of greatness. The output of persuasion it to have the decision maker take some action in your behalf. In order to do this, the thoughts and actions must be transplanted from the speaker into the decision maker. It is not sufficient to be eloquent; the verbal framework must enclose content of value.

Content is paramount. It is imperative that the wizard technologist makes transparent, understandable presentations that are rooted in sound content. All the style, techniques, and methodology are wasted if the content is weak. Ensure that the content, the facts and data, are true, validated, and solid. Then, and only then, work on delivery. Never purvey unsound technology.

Now, if the content is solid to begin with, choosing the right words will force the presenter to carefully examine the content. It will point out deficiencies and weaknesses in the technology, itself. In other words, selecting words carefully and articulating them well will not take away, or "spin" the content. Quite the contrary. Only after you frame the ideas with the proper words can you know whether the premise truly drives to the conclusion. Only by putting cogent, succinct words in place can you link the data to the recommendations and determine if the link is undeniable and valid.

Articulating the Concepts

- Add dynamics to the voice. Avoid monotones and mumbling.

- Language accents can be used as assets. The emphasis is on clarity, enunciation, and grammatical correctness.

- Avoid language that differentiates gender, race, or religion. Avoid profane language.

- Avoid acronyms and jargon.

- Learn the common grammar pitfalls and avoid them.

- Use the right word. Choose wisely.

- Words matter.

Chapter 10: Final Form
Putting It All Together

The lecturer should give his audience full reason to believe that all his
powers have been exerted for their pleasure and instruction.
— Michael Faraday

I drove to a guitar show in Arlington, Texas with my good friend, Dwayne Beck. Neither of us is a guitarist, except for a few chords here and there, but the show makes for a fun weekend and good company. Factory and custom guitar makers support the event, accompanied by hundreds of vendors setting up tables everywhere. We were popping hot dogs and roaming about when Dwayne created a notion that *we* could *make* an acoustic guitar. "If these guys can do it, why can't we?" he said. I could think of many reasons why we could not but I do cater to a challenge. And there was no hurry, no time frame, so we could take our time and learn as we went. But, how to start?

I had little to commend this pursuit. First, I was traveling with my business and time was lacking. Second, in terms of actually playing a guitar, I was limited to three-fret chords. If you want to sit around the campfire singing *Kum Ba Yah*, I'm your man, but limited beyond. Third, I simply did not know how to do it. *Aporeo*, again. To my credit, I had carved a respectable violin and I knew something about tools, but this would be different. Where to start? We bought a book on how to make acoustic guitars and both read it. It would be more difficult than I thought. (Dwayne never experienced such doubts.)

I researched types of wood and purchased the raw hardwoods for the back, sides, fingerboard, and neck. I bought tools for carving and sawing. I made forms and jigs for bending and gluing. I added scrapers, planes, sandpaper, and a spray gun. Soon the wood, tools, and parts arrived. I hung the tools on my garage pegboard and laid the wood on top of my work table. It sat there for at least a month.

It would all still be sitting there if my friend had not called one day and said, "It's time to cut wood."

That is where you are now. You have the techniques and tools laid before you. It is time to cut wood. It is time to be that persuasive wizard. It is time to shape yourself into a solid purveyor of technology, time to put a persuasive presentation together and deliver it correctly. If you follow the guidelines of this book, you will become a persuasive force for technology advancement.

SELECTING THE PEER REVIEW MEMBERS

You have been asked to make a technology presentation. You go through the steps outlined. You have all the components in place. It is time for the peer review. First, we define what is meant by a "peer" and what is to be accomplished in the review. It is probably different than your education and experience.

Should you enroll in the standard communications course, you will find yourself joining a class of like individuals, people who want to communicate. After some basic instructions, students give speeches to fellow classmates. Members critique each other's speeches. The instructor acts as moderator. This practice, while maybe helping with the basics of mannerisms, stage fright, and pointing out "and-uhs" is of limited help in technology persuasion. The students are learning communications. They are not knowledgeable technologists nor experienced decision makers.

First of all, such classmates base most comments on subjective supposition, not objective experience. The classroom text and the give-a-speech context is not geared to decision makers and technology, so the chain of ineffectiveness adds useless links.

When we talk about a peer review, here, we do not necessarily mean people you work with. The peers in the classroom were peers because they were enrolled in the same class, not because they were technologists knowledgeable of your specific area.

A candidate for your technology peer review needs to possess three characteristics: technical knowledge of the material you are presenting; knowledge about the decision you are requesting, its importance and criteria; and knowledge of the guidelines of this book. If the third element is lacking, you can teach it to them. Those are the persons who should constitute your peer review.

If some of the peers are in management, that is all the better, but not required. The purpose of this peer review is to ensure that the components of persuasion are present in the right amounts. You are not baking the cake, yet; you are ensuring you have all the ingredients in the proper amounts. In laboratory terms, you are not performing the experiment, but you are gathering the supplies.

PEER REVIEWS

When you are ready, gather a set of peers, about 3-5 knowledgeable persons who fit the three criteria we discussed. If possible, include in your audience a naysayer or two so you can

Perform peer reviews. Make modifications to the data, strategy, logic, key conclusions, and the graphic illustrations. Avoid text modifications.

ensure honest assessment and challenged probing. Do not simply gather people who worked on the project. Make them as broad as possible within the criteria. You are trying to determine five things in the peer review.

1. Does the presentation follow the *Why, What, How* scenario?
2. Do the data support the conclusions?

3. Is the connecting logic succinct and understandable? In other words, is the *What* component correct?
4. Do the data and logic compel the recommendations?
5. Is the balance right between the amount of *Why, What* and *How*?
6. Does the presentation length fit the time (using the time rules from Chapter Seven)?

Present your draft presentation to the peer review team. Do *not* practice beginning-to-end. This is not a practice in speech making. It is honing the data, logic, and strategy for persuasion. The peer review is answering the question, "Do you have the *components* of a persuasive presentation?" It is not "Do you *have* a persuasive presentation?" Go through the review in the order of the steps, above, not the order of the draft viewgraphs. Have someone take notes as you go along so you can focus on the review and not be stopping to jot down comments.

Does the presentation start with a *Why* component and end with a *How* component? Does it really? Separate the presentation into these components and show your audience of peers which part is which. Do they agree? Do you have the *Why* part right? Do you have the *How* part right? Does your final chart say what you want it to say? Does it make specific recommendations and tell the decision makers *How* they can implement them? Is it too matter-of-fact? Is it quantified or is it just a general statement of desire?

Review the main points in the *What* component and see how they play out. Does your presentation answer the natural questions that will be posed by the decision makers? Does your knowledgeable peer audience feel this information is ready for "prime time?" Are there errors or voids in the data? If so, what needs to be done to fix that? Is the logic in the right order? Does the logic flow well? Is the logic forceful? Is the logic unambiguous? Is there logic?

Is the proportion of the *Why, What, How* about right? This proportion should be, nominally, 10%, 75%, 15%. The most likely culprit are the *Why* and the *How* components. The *Why* is usually much too long. The *How* is usually a summary and much too short. Make sure that all three components, *Why, What,* and *How,* are succinct and concise.

Now, look at the bulleted text in the viewgraphs. Is the presentation full of negatives? It is common to get to this late point only to realize how negative the presentation sounds. "We do not have this. We cannot work without that. We're unhappy with this. We can't get it done on time. We don't have enough money. We don't have enough people." And so on. Instead of saying that you do not have the funds promised, describe what you would do if a specific amount of funds were available by a specific date. Or, alternatively, what you must cancel because the funds are not available. After the Peer Review, take time to rewrite these text bullets in positive form. During the Peer Review, just mark them for later modification.

This is also the time to replace the vague, ambiguous, or non-quantified statements with specifics. Remove meaningless adjectives—like almost, seldom, usually, sometimes, nearly, lots of, very, best—and replace them with quantified values. If you cannot quantify them, it is usually best to leave off the adjectives entirely.

Remove all "spin" and replace it with objectivity. To the extent possible, make the presentation an unbiased assessment.

Please note: other than converting negative statements to positive ones, we have said *nothing whatsoever* about actually editing the bulleted text in the draft. In the peer review, look at content, intent, and logic. Do *not* waste time on text. Unfortunately, what most participants love to do in these peer reviews is "wordsmith," a coined term—they all become grammar queens. They all want to take each point, each bullet, and work on the wording. Then someone else will make a modification to that, and someone else will disagree, and someone else will want it back like it was, and soon you have an entire committee editing every word, every bullet. It is much too early for that and it gets nowhere. Let them wordsmith *only* the very first viewgraph (the one after the title viewgraph) and the last viewgraph: only these two. For the rest of the viewgraphs, set the guidelines such that they may critique only content and meaning. Tell them you will work on the wording, offline. You will have to go through this litany of reminding them several times, and perhaps many times. Keep at it. Wordsmithing is a tough habit to break. This is not the time for it. It is counterproductive at this point. You do not even know if you have the components yet.

We held a review for remodeling the entrance of a new office addition. The architect laid out the drawings. There lay the electrical plans, the plumbing, the air conditioning, the floor layout, the furniture, the perspectives, the cross sections, everything. He previewed a large color sketch of the front entrance. For the main door, he had selected a distinctive doorknob. One individual remarked on the knob as perhaps too ornate. Another person agreed. Another disagreed. Still a fourth had an even different idea. It kept going. More designs and opinions. I looked at the clock—twenty minutes and we were still envisioning a doorknob. A doorknob mind you. Here were all these drawings, all this technology, and we were arguing the doorknob. Why? Because everyone has an *opinion* on doorknobs. They have no facts, data, or information, just opinions, preferences, and notions. The electrical and the plumbing are boring and difficult. Those require facts you expect someone else to verify. Everyone knows what a doorknob looks like. You must take command in these peer reviews. They rapidly degenerate into designing doorknobs.

Wordsmithing, in the first peer review, is a doorknob. Words are essential. Words are important. Words are critical. But first you must decide *what* you want to say before you spend all your time on *how* to say it. You must do the hard part first: examine the data, logic, and recommendations and quantify all of it. You will have to work hard to constantly steer them away from wordsmithing. Concentrate on the content.

Refine and hone the Visual *Aides*. There should be only two or three of these. Remember, Visual *Aides* are not ordinary viewgraphs with just pictorials and graphs. These are the stand-alone thought generators, the *Aides*. Read Chapter Eight if you are uncertain. These are the visualizations that transcribe the logic and that drive to the conclusions. Are they effective? Is the impact immediate? Ensure they are not trite attempts at humor or summarization. Humor must be spontaneous. A summary is unnecessary.

Finally, look at the time allotted against the material laid out in front of you. Using the time rules of Chapter Seven, can you give this presentation in the time allotted? Remember, for an hour's scheduled presentation, you have only about 40 minutes of actual presentation content. The rest of the time is used in starting and stopping, interruptions, questions, and answers, all controlled by the decision makers, not you. Almost all technology presentations are too long overall. In particular, they are too long in the beginning and too short at the end. If you think yours will fit into the time, look again. Remove redundancies. Are there elements that simply should be deleted? Work on compression. Go back to your *Spider Diagram* and review the priorities you delineated there. Does your presentation follow that priority? If not, why did you change it? Are there things you included that are actually "below the line" in priority? Is there a summary of any type? If so, remove it.

In the end, you want your peer review to determine if this presentation is complete, succinct, and effective. Does it get the message across? Does it have a message? Is there sufficient technology to back the statements? Is it honest? Does it give specific recommendations? Are those recommendations achievable and realistic? Is it accurate? Is the case for adoption strong or weak? If weak, what can be done about it? If strong, is it too dogmatic, too demanding in tone? Are there additional data that must be collected?

What is missing? is there content that is confusing? If so, how do you correct that? Are there superfluous things? If so, eliminate them. Too many details? (Rarely does a technology presentation have a *lack* of details.) Does it have unnecessarily eristic elements, self-serving components, a lack of quantification, or a lack of substantiating data? If so, repair those deficiencies or put plans in place to so do.

When you complete the peer review, you should come away with a full set of notes taken by the note taker. Take a short break, but as soon as possible, start to work on implementing the changes from the peer review. Ideas age rapidly. If you wait until the next day, you will lose the feeling and emphasis of many of the comments. The next day, you will forget what some of them even meant, or why they were important.

AFTER THE PEER REVIEW

In *The Adventure of the Abbey Grange*, Sherlock Holmes examines the evidence, assimilates the clues, muses potential connections, and suddenly he has it. He cradles his mackintosh and shouts across to Watson, *"Come, Watson, come! The game is afoot! Not a word! Into your clothes and come!"*

This persuasion game's afoot. You are ready to coalesce all the material. If you had the Peer Review that is typical, you have much to do. First, you should *not* make all the changes recommended by the peers—you are not an editor, you are the persuasive wizard. Review all the recommendations and all the notes taken by your note taker. Evaluate each. Do, indeed, make the changes that are logical, necessary and effective. Make sure the changes fit the techniques and criteria of this book. Make those changes and put them together in a second draft.

Now, and only now, look at the bulleted text. I recommend this not be done in a peer review for reasons I discussed earlier. Remove the buzz words and acronyms that have crept in. Look again for any negative statements and turn them into positive ones. Defuse any land mines, those unnecessarily controversial comments that may open up Pandora's Box. You do not want the decision makers to be exploding shrapnel charges. Remove any potentially political, social, religious, gender or ethnic statements that might be considered offensive. Choose wording that is unisex.

Look and ensure there are no soapboxes hidden in the presentation. If so, remove them. A soapbox is a favorite theme that, regardless of the context, you imbed somehow. It can be a new laboratory, more equipment, a salary increase, more personnel, or whatever. This is a presentation designed to persuade. Those outworn, peripheral issues will tarnish the wares. For example, I had an engineer who felt unappreciated. So, no matter what the subject, he worked in how much he contributed and how he should be rewarded. Whether ostensively, discretely, or subconsciously, he rang the same bell. Dismantle the soapboxes.

Simplify the graphs and charts. Label all the graphs, plots, photographs, sketches, and drawings. Make them clear, precise, and font-readable. Ensure that the graphs, charts, and pictures are cogent and not just thrown in for filler and fluff. Throw away all clip art unless you can make a strong case for its efficacy. Graphs, plots, photographs, sketches, and drawings should convey a strong message, not just fill white space or add color.

Work some more on the Visual *Aides*. They covet and reward attention.

Look for potential copyright infringement. While it is common to extract from the internet and insert into a presentation—texts, graphics, photos, and the like—it is likely a violation of copyright law. Even if the material on the internet is not marked as copyrighted, it is still protected. In addition, why would you want to plagiarize? Be original. Public domain items are acceptable, but ensure they are not ineffective fillers.

Ensure that the grammar is correct. Check the spelling and the grammatical construct. Most technologists are weak in this area. Do not hesitate to ask for help. Edit to ensure consistent fonts, styles, and other standards or guidelines. Get a technical publications expert to do this if you really want it right.

Now, look at your completed version. Review, again, the notes from the peer review and check them against what you have done. Ensure you have followed the guidelines of this book. If so, you are ready to go to the Pink Team or Red Team Review.

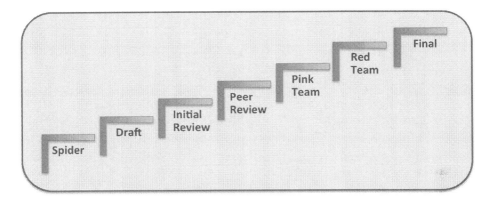

Figure 10-1: Process for Honing the Presentation. Ensure that each step in the process adds unique value. Each step has a different focus and intent.

WHY NOT PRACTICE END-TO-END IN THE PEER REVIEW?

It is a fair question to ask, "Why not practice the entire presentation end-to-end at the peer review? Is not that what it is about? Why do I work on it in pieces?" The answer is because practicing the entire presentation, to this point, is counterproductive. We are working, now, on the components and the techniques.

Consider a pianist. No student practices a new sonata by playing it beginning-to-end, beginning-to-end, time after time. No teacher teaches that method of practice because it does not work. The entire piece is practiced bit by bit within each section. If the student finds a troublesome spot in any *part* of a section, he practices that part over and over, over and over, until it is perfect. He does not start back at the beginning of that section, but stays right on the problem area until it is perfect. He practices just the part that is causing the problem. He trains the mind to do it correctly. If he goes back to the beginning every time and tries to go straight through it, he simply accumulates mistakes. He trains the mind to make mistakes. Do not train your mind to make mistakes.

Consider baseball. The pitcher that starts the game is called the "starter" and is expected to pitch six or seven innings. A typical starter can do extremely well for three or four innings. He may keep the opposition team to no runs. However, by the fifth or sixth inning, he will have pitched 70-90 pitches. The opposition team has gone through its rotation of batters a time or two. By the fifth inning, every batter has seen what this pitcher has to offer, his slider, sinker, curveball, and fastball. The starter is getting tired and all the batters are getting wiser. Faults now show up in the starter that did not appear earlier. Batters are watching for the "tell." The starter makes mistakes. The batters begin to score. Home runs are hit. We are losing the game.

The "closer" is the pitcher who pitches only one inning, the ninth, and then only if the team is winning by a run or two. He is a very good pitcher, possibly the best on the team, but he cannot maintain his great pitching for enough innings to be a "starter."

In between the starter and the closer is the "bullpen," composed of pitchers who pitch, nominally, the seventh and eighth innings. If the starting pitcher flags early and too many runs are scored, then the bullpen comes in early to rescue the starter. Occasionally, the starter does not even make it through the first inning. Too many hits occur and the bullpen is called to duty.

The technology presenter must be all of these: the starter, the bullpen, and the closer. It is difficult, as all wizards will testify. Most of the game is played in the first eight innings, not the ninth. If you do not pitch well in those first eight innings, you do not have to worry about closing. Many wizards make reasonable technology presentations when they do all the talking, when they are totally in control. They become effete when a question is asked or controversy arises. Challenge is difficult. Similarly, many wizards have no ability to close. They come to the end of their spiel, peter out, and wonder what to do next. Presenting to decision makers requires the full cadre of skills and techniques.

This idea of practicing the presentation beginning-to-end probably comes from classroom training. In the communications class, the entire speech is nominally 5-10 minutes, total time. They are this short because of the class size and class time limit. With 10-20 students in the class, it is difficult to have time for even one long speech during the entire year. The 5-minute time limit does not develop components of the technology presentation. It is more like the elevator speech we discussed earlier. The 5-minute speeches, the elevator speeches, serve somewhat in the office of a closer— quick, succinct, and to the point. This "closer" training is important and necessary, but scarcely sufficient. Training with elevator speeches will not develop the logic, reference, and reach-back skills that are needed for the marathon decision-making presentation. The peer review isolates these elements—starter, bullpen, and closer—so you can perfect each in turn.

PINK TEAMS AND RED TEAMS

If the ultimate decision maker is several layers above you in the management chain, then you will get the "opportunity" to review your presentation to the in-between layers of management. The purpose of these pre-reviews is to "get it right." They do not want you to embarrass you or them in front of the Big Guy. These preliminary, mostly

> *Dry-run the presentation with a knowledgeable team, the so-called Pink or Red Teams. Look for holes in the logic first. Make text changes last. Convert all negative statements to positive statements.*

management, reviews are commonly called Pink Team or Red Team Reviews. Thus, a Pink Team Review and a Red Team Review are simply practices prior to the final presentation.

A Pink Team or Red Team is composed of individuals who are knowledgeable about your final targeted audience of decision makers. Most, but not all, will have limited knowledge of the specific technology. (The peer review focused on the technology.) The Pink and Red Team members roleplay, so to speak, the final audience. But, they are very good role players because they know the key decision makers intimately, or at least have considerable experience with them.

Both the Pink Team and the Red Team can have technology members present. Usually, these are technology managers, but it is not uncommon to bring in a technology consultant or other technology "expert." The emphasis will still be on preparation for the final meeting and not on minute details of the technology.

The terms "Pink" and "Red" are descriptive of their functions. Red is a color that screams, "Alert! Get ready!" Thus, the Red Team Review is the final review before the actual presentation to the Grand Poobah. Sometimes, the Red Team members are so high-ranking that you need a lower-level Pink Team review before you even get to present to the Red Team. Anyway, the colors are just arbitrary to indicate order, and pink comes before red. You may have one, or both, or neither.

Whether the review gets called Pink, Red, or just a review, this is the one time when you do, indeed, give the presentation. It is not like a play rehearsal, though, where the lines are memorized and you go through it verbatim. You get interrupted many times with helpful comments and questions. You eventually work your way through the entire presentation but it is not smooth. In the Pink Team or Red Team Review, you may not say every word, but you do go through the entire set of viewgraphs and the *gist* of your words.

Hopefully, at this review, you are making only small changes. Since they are not technologists, most of the Red Team members will recommend changes to the words on the viewgraphs, or the words you add verbally. This is extremely important, now, because they know the mind of the key decision makers. Give heavy weight to their comments. Again, they will gravitate to wordsmithing. However, ensure they examine the logical steps and the content. Use this opportunity to get support for your recommendations. If there are any on the Red Team who are opposed to your conclusions and recommendations, be sure to discover why. If at all possible, try to get them on board with your idea. You do not need naysayers, especially naysayers who may adversely influence the key decision maker.

In many cases, the Red Team finds significant issues with your presentation. The error is usually not in technology. Wizards are good at technology. It is usually that the technologist did not follow the guidelines of this book. The issues are generally outside the technology arena. Follow these guidelines and you will have no problem with the Red Team.

If the Red Team issues are fatal, the final meeting should be cancelled or rescheduled, if possible. In most cases, however, the final meeting is fixed in time and you have to go to work to repair the shortcomings. You may be required to appear again before the Red Team. It depends on how egregious the problems, how important the decision, and how elevated the final decision maker.

If there is no formal Red Team, then form your own and treat it as you would a formal review. You need this opportunity to present your entire presentation in the context of the intended audience. Do *not* solicit peers and have them roleplay. What possible knowledge do they have of how business-world decision makers think? Plus, they are too close to the proposed recommendations. Get independent persons who are familiar with the ultimate decision maker, and who have not participated in the technology work itself.

Why? Because the audience of peers has no idea how decision makers in the real business world make technology decisions. Preconceived notions of what a decision maker is like are erroneous. It is analogous to a concert pianist preparing for her concert by performing a Bach concerto to a group of high school students. With little exposure to classical music and a yen for rap, how useful is their critique, really? Most of their comments would actually be *counter*productive. And, there would be no way to calibrate the validity of any of the comments. Such is the case with having a technology peer pretend to be a decision maker. Again, do not use peers or senior technologists if you create your own Red Team review.

FINISH

Review the changes recommended by the Red (or Pink) Team. Consider their recommendations carefully. Make all changes that are required. If other changes were just "suggested," consider them carefully. Make all changes that will get you the decision you want.

> *Edit the text and finish. Perform spelling check, correct grammar (number, tense, and similar), ensure format consistency, and adhere to policy convention and copyright permissions where applicable. Make final copies.*

The Red Team almost never has a comment on the actual technology. However, if there was an expert or outside consultant on the Team, then you may be required to get more data. Usually, though, the Red Team finds fault with the presentation content, logic, or conclusions. Unfailingly, there is too much data, too many details, and not a strong enough case for *How*. Think in terms of the final decision maker and her criteria. Think of the end goal and not all the data in between. Always keep the end goal right in front of you.

Make all the changes. Perform a spelling check. Edit the viewgraph text and then finish. If possible, put it aside for a day or two, for at least a couple of hours. Then, go back and review it when you are fresh. Make final changes. Do not be surprised if you make considerable changes. It is amazing how different things appear when given time to marinate.

Finishing Up

- Carefully select members for the peer review.
- Review the presentation components in the order prescribed.
- Focus on the data, the logic, the conclusions, and the recommendations.
- Emphasize the Visual *Aides*.
- Carefully select members for a Red Team. These should be persons knowledgeable about the final decision maker.
- Perform a final Red Team review.
- Edit the text and finish.
- After several days have passed, review the presentation, again, and make final changes.

Chapter 11: Loaded Deck
Leveraging For Results

δοσ μοι που στω κα ικινω τη νγην
Give me a place to stand, and I shall move the earth.
— Archimedes, after discovering the principle of the lever
As quoted by Pappus of Alexandria, ca. AD 340

We create and develop content. We ensure that the data and conclusions point forcefully and uniquely to the recommendations. We analyze the components in a Peer Review, refining the technology, honing the logic, and sharpening the conclusions. We review with a Red Team to target specific decision makers. Is there more? Yes. The job is persuasion. It requires more than just *having* the goods. You must *deliver* them. When you get into the conference room, you can posture and posit your arguments in a manner that strongly leverages the results. That involves some strategies and techniques that move the pivot point toward success.

NO APOLOGIES, PLEASE

It is intimidating for the typical engineer or scientist to enter a room full of senior leaders, entrepreneurs, or other decision makers. There is an inclination to self-reduction. In spite of your "youth and inexperience" (a quote from Ronald Reagan, age 73, about his presidential opponent, Walter Mondale, age 56), in spite of the circumstances, never begin the presentation with an apology. Do not apologize for any inadequacies of self, nor for your preparation, your lack of time, or lack of skills as a speaker—even and maybe, especially, if they are true. Do not apologize for your shortcomings, your nervousness, or your lack of familiarity with the subject. Do not do this. It accomplishes nothing. Make no apologies. You are what you are and you have what you have. Any shortcomings will be evident quickly enough. The decision makers have been here before. Do not overtly point out your weaknesses or try to paint them with what will appear to be excuses. You are there for a purpose, so begin with purpose and you may find yourself more capable than you imagined. If you are not, any apology will little ameliorate the deficiencies.

To begin with, an apology brings the conjecture to the mind of the decision makers, "Then why are you here?" Suppose you say, "I'm sorry, but I'm not the most knowledgeable person here regarding this subject." They may think, "Then why didn't you become knowledgeable or bring someone else who *does* know the subject?" You lose. Suppose you say, "I had insufficient time to acquire and analyze all the data I needed." They think, "Why didn't you work harder to get more data?" You lose. There is no good answer to any of these questions. All your inadequacies may be true, and they may not be your fault, but who will care? To bring them up and challenge the audience with your inadequacies is self-defeating. Whether your presentation is mincemeat or sirloin, make no apologies. Accept the outcome, whatever it is, and fashion it to improve.

I understand the psychology. In the mind of the novice, the apology somehow makes up for lack of data or lack of ability and gains sympathy from the audience. You want the decision makers to overlook certain inadequacies with statements like, "I'm the new kid on the block," or "I'm not accustomed to presenting at this level," or "I received this assignment just last week and have not had sufficient time to get the right data." This does not, however, produce the sympathy, compassion, or understanding you desire. At best, it shifts the blame to someone else, who is probably in that very room, wishing bad things.

Decision makers are busy, important people. You just told them that they are not important enough to receive the presentation from the "right" person. Start with an apology and you inform them that "someone" made a big "mistake" and they are wasting their time listening to you. You will not get sympathy, you will get annoyance. At best, you receive tolerance.

If there are inadequacies in the data, quantify these when you reference them later in the presentation. At that point, those inadequacies will be apropos to the conclusions and recommendations, not an excuse. You can show how the lack of, or inadequacy of, the data affected your conclusions and subsequent recommendations. Quantify any shortcomings in your analysis, yes indeed, but do it professionally and not as an apology, and certainly not at the very beginning.

William F. Buckley, Jr. was known for his conservative witticisms, quick logic, and sesquipedalian words, a wizard of another genre. He was a student of the harpsichord. Because of his notoriety as a political pundit, the Phoenix Symphony Orchestra asked if he would consider playing a Bach Concerto with them. Buckley had limited skills as a harpsichordist. "For sheer love of Bach," he consented. Buckley came on stage knowing he was out of his league, knowing he was not a professional, being noticeably and affectedly nervous. After his performance, the audience gave him a standing ovation and he even performed a credible encore. The next morning, the critics were more realistic, "Too many wrong notes to ignore." "Highly nervous." "He and the orchestra at least managed to finish at the same time." And so on.

Was he a professional musician? No. As measured by professional musicians, did he do well? Probably not. Was he chagrined? Yes. Did he apologize? No. You are what you are and you have what you have. The audience knew his lack of credentials and they chose to hear him even above his inadequacies. He did not advertise them. Do not advertise yours.

If you are chosen to present the information, then present. Take whatever invectives and criticisms are hurled, and use those to improve. But, do not try to lessen the expectations or ameliorate the pain by apologizing. It is what it is. Apologies are excuses. Decision makers are not looking for excuses or apologies, they are looking for results. So use the experience to produce better results the next time. Losing a single battle is not losing the war. Now, on that particular day, at that particular time and place, it may seem like the war is lost, but it is not. You are the persuasive wizard. You will not make the same mistakes twice.

LAST AND FIRST WORDS

Of all the viewgraphs you show, the last one and the first one are the most important. The content of these may determine success or failure. The first will set the course. The last will stay in their minds when they leave the room. Make these viewgraphs and your accompanying words rise to the occasion.

> *Memorize and practice your opening and closing lines.*

As you prepare for the meeting, lay, on a table, the first viewgraph (the one after the title and topics charts, the first real viewgraph) and the last viewgraph. The first should say *Why* you are there, what you plan to decide, what you plan to accomplish. The last viewgraph should answer the questions posed in the first viewgraph. By no means should the last viewgraph be a summary or a *Te Deum* of acknowledgements. The first viewgraph is the *Why*. The last viewgraph is the *How*. The last viewgraph is the implementation. It tells the decision makers how to improve the world by enabling your recommendations.

Take extra time with this last viewgraph. Just as the words of a dying person have significance, the last viewgraph is most significant. (An apropos analogy.) Carefully check and recheck the wording of the last viewgraph. Make it positive and dynamic. It can be words, a graph, a chart, a picture, or whatever. Avoid cartoons, trivia, and trite statements. This is your "walk-off-the-stage chart." Just as you would flip the light switch when you leave the room, you want this last viewgraph and your words to toggle their decision in your direction. Most opinions are disproportionately weighted by the final input, the last thing heard, the last thing seen, or the last thought. Make it count.

The last viewgraph should close the deal. Make sure it does just that.

The first viewgraph should open the deal. Make sure it does just that.

I do not recommend memorizing the presentation. This would be noticeable, boring, and contrived. However, I do *require* that you memorize your first lines and your last lines. Prior to the meeting, memorize the first few lines you want to say to the audience and the last few lines. There are several reasons for this. First, you will be nervous when you start. Knowing what you are going to say at the very beginning will get you started. It will also ensure that you start off in the right direction.

Second, by the time you get to the end of the presentation, your head will be full. You need to have rehearsed how to make your exit. In addition, and thirdly, these last few lines must close the deal. Make them surgical, possibly pithy. This is your last chance. Fourth, reciting your memorized final lines lets *you, yourself,* know that you are finished. Stop and get off the stage. This ability to finish well is one of the most difficult, yet most important parts of the presentation. You have spoken well, you have made significant points. How do you stop? If you do not prepare in advance, if you try to come to a stop in real time, it is hard to avoid that feeling of either draining to a standstill or being jerked to a stop by a frustrated audience. If you prepare and memorize in advance, you start with energy and finish with élan, you start with direction and end with decision.

That first chart, make it grab their attention with *Why.* That last chart, make it count and close the deal with *How.*

DEMEANOR

The right demeanor, or attitude, is critical to success. Deliver with integrity, professionalism, and acumen. Be courteous to all, receptive to counter-arguments, gentle with controversy, and responsive to suggestions.

A technology presentation to decision makers can be tricky because the audience usually is either not technical or only weakly technical. Even so-called high-tech companies are being run more and more by persons with management, business, investment, or other backgrounds, possessing little or no formal technology education. You, the presenter, must fill the gap—but it can be problematic.

The most common demeanor problem is that of being too techie, a geek, spouting acronyms and flipping buzz-words. After poor performances, amateur presenters told me they thought the audience would know all those terms. The audience did not. This constant insertion of unfamiliar jargon strangles communication as the non-technical audience attempts to piece together the segments they do understand with those they do not. If they cannot do this *readily*, major points are choked out and success is extirpated.

Another potential demeanor problem can occur. It is common for a technology wizard to waft an air of superiority, being arrogant or even haughty. It may be subconscious. No matter.

I was with a group of software users. A former chief engineer from Apple, Inc. was making a presentation about his company's product, an online cloud application connecting third-world users with specially-selected US counterparts. One of the audience members asked about security, how this might affect her personal email account. The engineer replied, "It is all very technical and you would not want me to go into that," whereby he proceeded to give a benign response that was naked of any technology that might confuse her. A second question came up about the use of different routers. The response was similar. "It involves a lot of technology you don't want me to go into." Quite the opposite, we did want him to go into some technology, but he hid behind his superiority thinking he was doing the right thing. Every audience resents being "talked down to." This pinnacle of assumed superiority short-circuits recommendations, no matter the wherewithal or capacity of either presenter or audience.

Ironically, this purported superior attitude can be a cover for a weak presentation. Like a defense attorney, if the facts are against your client, then argue the law. If the presentation is weak, then many wizards argue their superior knowledge. It is undoubtedly more effective in the courtroom than in the boardroom. Invariably, it falls flat. Do not go there.

If not due to weak material, then this foisted superiority can be the sign of an inflated ego. Or, it can be a compensation to the natural introversion most technologists feel. Whatever the cause, the audience of decision makers has no problem with diagnosis and the general prognosis is failure to approve the recommendations. The audience may leave the room remarking how brilliant you (probably) are but you will not get the decision you sought. They will not enact the things you recommend. All the data and energy are wasted and ineffective if you let *yourself* be your focus.

When a technical presenter presents on too simple a level it leaves the impression of superiority and talking down to the audience. It may be unintentional, but it is counterproductive, nonetheless. It resets the listener from attention to affront, from content to intent. It is one thing to think (incorrectly) that all decision makers are ignorant of technical subjects, but it is a bigger mistake to *act* as though they were. Those decision makers may have finance, legal, political science, history, or other non-technology degrees, but in the hundreds of presentations they have sat through, thousands of documents they have read, those decision makers are not dummies. Often, they know a lot about technology, even if they are not, themselves, firsthand experts. They may know far more about your subject than you do of theirs. But, regardless of their technical training or lack thereof, they will not condone intellectual subjugation. Respect all audience members as intellectual peers.

Another demeanor problem can arise if one of the decision makers chooses to bring her own technical expert to the meeting. Decision makers universally and invariably believe that "the geniuses work somewhere else." They have trouble accepting that the expert they desire may be the one in front of them. "A prophet has honor except in his own hometown." Sometimes, these experts are outside consultants, sometimes in-house experts. Always, it is problematic, not because they find shortcomings in the presentation or content, but because they bring their own agendas that feature a need to show their own value to the decision maker. That usually has them target at least something in your presentation. Suffice it here to say that your attitude must be one of respect while you remain firm in your convictions. You did not come to your technical decisions without hard-gained insight, so fight for what you believe to be correct, but be a professional about it and show respect for counter-opinions.

In most cases, you, the speaker, are indeed the most technically informed in the room, at least about that particular subject, on that particular day. You are the expert. So, act like one. Never, never, never, under any circumstances, condemn or belittle anyone in the audience. As the presenter, you are in some position of dominance because it is awkward for them to retort. Maintain authority with dignity and professionalism and never resort to arrogance or put-downs.

BE BOLD

Boldness, based on knowledge and education, facts and data, is a strong suite. When asked by the president of the company, "Based on your data and your experience, do you genuinely believe we should continue with this technology," you must be able to commit. If you are vacillating and irresolute, that uncertainty will undermine your entire thesis, regardless of its merits otherwise. If the data do not support the recommendations, then, by all means, do not make the recommendations. But, if your data and your knowledge and your experience point to a recommendation, stick to your guns.

Our legal department was examining a potential patent infringement. Had the competition perpetrated our intellectual property? The attorney's question was direct. "Did this other company infringe upon our patent?" It was both a technology and a legal question. He knew the legal aspect, but he was asking me about the technology. "Is there a violation?" He wanted a technology answer. He was not the technology expert, I was. He did not want qualifiers like, "Maybe," or "I think so." He wanted verifiable commitment, "Yes" or "No." You must stick to your decision and live with your response.

This is not to say that you are bold in the face of uncertainty or lack of substantiation, but if the results justify the conclusions, be bold. If you are weak and vacillating with your very own technology, how can you expect decision makers to be bold and commit to it?

No project is one-hundred percent certain. When there is uncertainty, quantify that uncertainty. You answer the president boldly, armed with facts, "The statistics give a 90% confidence, which is sufficient to make this recommendation." If greater certainty is required, you can remark something to the effect, "We believe we can achieve a 98% confidence factor within six months at an estimated cost of $350,000." That is the type of boldness the decision makers desire. They want an expert opinion they can understand, quantified in terms they recognize. Maybe it is not boldness, exactly, but it is a verifiable, quantifiable, confident expectation. Of course, you must be able to explain to their satisfaction and understanding how all this can be.

Wizards drive decision makers to drink when they say, "This is *research*. I don't know when we will have sufficient data to make a decision, nor what the conclusion will be, nor how many people it will take, nor what it will cost—but I'll let you know when I'm finished." Make these statements and the decision makers will rightly conclude that it never will be finished. Scientists love to hide behind the phrase, "It's difficult to predict, especially research." It would do Yogi Berra proud.

Say the owner of the Boston Celtics asks the coach, "Will we win the Eastern Conference Championship this season?" Should the coach respond, "I don't know, it's difficult to predict." Or, should he say, "We'll just have to play the games, one at a time, and see what happens. I'll let you know after we're finished." Huh? Would you run a business this silly way? No, and that same silliness inflames decision makers when they hear, "It's research. I can't predict when I will be finished or what the outcome will be."

Now, the owner of the Celtics does not want the frivolous response, "Of course, we'll win. Go team! Rah! Rah! Rah! Sis! Boom! Bah!" Here, again, is silliness. What the sports owner, the decision maker here, wants is a breakdown of what the coach is doing, or what can be done, to ensure the win. What is the physical condition of the team? What is the strategy? What is the logic the coach is using to arrive at his predictions? What changes can be made to ensure a win? New players? Which players? Higher salaries? How high? Which ones? New stadium? Why? What might be the impact of these changes? What *decisions* need to be made? What needs to be funded and to what level?

Likewise, the technologist, or scientist, must make predictions based on experience and knowledge. A wizard will never know everything she would like to know. She will never know everything she even needs to know. That is what makes experience and information so valuable. Take the data and make solid conclusions and then stick to your certifications. Quantify all uncertainties. This is not the time to be self-deprecating or make apologies. Decision makers want someone who knows the subject and can give them answers that are right most of the time. You are allowed a few mistakes, but you want to develop a reputation for being a solid thinker and correct a large percentage of the time. Boldness is humility wrapped around a firm foundation of knowledge, experience, data, and analyses.

BODY LANGUAGE

If your plan is to put viewgraphs under the noses of decision makers, you can do that with a monkey. The appearance, poise, and attitude you project reinforces what you say, what you show, and what you know.

I listen to music. I listen to CD's of my favorite groups. I watch videos of their performance. I like some of them so well that I fork over hundreds of dollars to hear them perform in person. Was there a difference between the live performance and the videos? Was Bingo a dog? There was no comparison, none whatsoever. This is the same effect your presentation can have when you put yourself into it.

Your body language needs to identify with your presentation. You need to believe in what you are doing. Here are some points on how to convey the message.

Smile: It is amazing the calming effect this will have on *you*. The tension and stage fright will melt. When you smile, your face tells your body, this might be fun, I might like this. It also puts the audience at ease by showing them the confidence you have. Make the smile the first thing you do before you ever speak and the last thing when you walk off the stage. Think about doing it in-between. Smile as you walk to the platform and turn to your audience. It is not important whether they smile back. The positive effect is for you as much as for them.

Facial Expressions: You cannot be persuasive if your expression is reminiscent of a Botox treatment. Express your thoughts with your face: not overdone histrionics, but natural emotions.

Arms: Move those limbs naturally and use them to emphasize the presentation content. One professional speaker put it this way, "Let your armpits breathe," and she held her arms out to demonstrate. What she meant was to let your gesticulations and hand movements be broad and expansive. Gluing your arms to your ribcage or nailing your hands to the podium reads novice, starchy, or nervous. In a technology presentation, there are many opportunities to shape something in empty space—from the operation of a gyroscope to the splintering of timber, from Buckyballs to bionics. When arms and hands come into play, you bring the audience away from the view-graphs, away from the conference table, away from the email devices, and back to you, and the central thought of your presentation. In doing this, you lighten their load. You create and illustrate with your hands.

I have known only one engineer who was so stiff as to simulate little R2D2. Fortunately, such personalities are a rarity. Most persons have a natural rhythm that just needs to be released during the presentation. Nervousness or tension often locks up the natural rhythm. Release it.

Body language adds leverage to your words. In Rome, I boarded the overly crowded subway train to the Coliseum. Two young women pushed me aside and forced their way to the center of the car, dragging shopping bags and purses, and never missing a beat of their voluble conversation. Each grasped the center pole with one hand and held their packages and purse with the other. We all were packed like sardines when the train jolted its start down the track. I watched these two persons chat. Their strapped shoulders noticeably shrugged in marked emphasis with their words. That evening, in the hotel, I switched on the local news. The usual talking heads in a US program were replaced, in Rome, by talking heads with syncopated shoulders. Words were not enough. They needed movement to communicate.

Different Strokes: In a foreign country, know body language and understand there are differences. In the Middle East, for example, you never want to show anyone the bottom of your shoe. It is considered an insult. Likewise, you never want to extend your left hand to take or receive anything, even in gesture. The left hand is reserved for the toilet paper. In Asia, we had a senior consultant with us who was in the habit of making the "OK" sign with his thumb and forefinger. That habit is a vulgar no-no there.

Here are some examples of how to use body language to your advantage. Point with your arm and finger to the viewgraphs. Let those armpits breathe. Do not point to the audience, though. Pointing towards someone, even at a distance, appears confrontational. If there are two or three key essential elements to be discussed, count them on your fingers as you elicit each one. If there are two options to discuss, motion with the right hand to discuss the one, and with the left to discuss the other.

Move About: Stand away from the podium and move occasionally. Change directions. If there are two sides to the argument, maybe you stand to one side of the podium while you discuss one and move to the other to discuss the other (standing in front of the podium at all times, of course). Move to different parts of the platform, but do it intentionally and with purpose, not in nervous pacing. If there is a break in the thought, then that is a good time to move slowly and smoothly to another spot on the platform, or change standing positions. Move when there is a natural break or pause in the logic. Keep speaking without hesitation while you move to the other location. Do not get in the habit of just moving around for the sake of movement or working off energy. This makes a speaker look like a novice and erratic. Make your movements intentional while avoiding any sense of practiced choreography.

If the room has obstacles such as supporting pillars, many persons will not be able to see you at every location. Changing positions gives everyone a chance to see the speaker, especially if David is sitting behind Goliath.

Most often, the physical movements are side-to-side. But if you want the audience to retreat from an option, then maybe you step backwards to reinforce what you are recommending. If you want them to advance, perhaps you should move forward. Most of these movements come naturally, but you have to do them. One has to avoid the extremes of never moving, always pacing, or making quick, abrupt changes. It takes very little actual movement to get the point across.

Eye Contact: Eye contact is essential. I was in Salt Lake City to meet with a scientist of a small company owned by him and his brother. The scientist brother began to explain his inventions. Only twice in an hour's presentation did he look at anything other than the white laces of his tennis shoes. He was most intelligent, yet most ineffective.

As you work through your presentation, look at individuals in the audience. Get eye contact. It was in a large group, but when I was with former president George W. Bush, I thought he looked right at me for concurrence. Maybe he did or maybe he did not, but it was effective nonetheless. Let each person in the audience know you are talking personally to him or her. Do not be shifty and do not stare. Smile occasionally as if you were enjoying the presentation. Comfortably look at members of the audience as you speak. Particularly, look at the main decision maker, but do not make the presentation just to that individual or the others will feel left out—it may take all of them to get your recommendations approved.

Read the Body Language of the Audience: The audience reads your body language. Learn to read theirs. They are constantly giving you feedback, even if none of it is verbal. Use their feedback to constantly, continuously, modify your presentation style. If one of them keeps looking at his watch, you are probably going too slowly. If the main decision makers are flipping through your presentation and looking ahead, then you are presenting too slowly. Reduce the details and begin anew to get their attention.

Wireless Devices: In most meetings, wireless distractions will be rampant. If everyone seems to be working on his wireless device, you are probably boring them. Speeding up may help. Think about what might be wrong and see if you can fix it, *in situ.* This is not always possible because it may just be a poor presentation and you cannot fix that while you stand there boring them. Put some inflections in your voice. Work on the pauses. Smile. Use more hand movements. Let those armpits breathe. Change positions on the platform. If the content is weak, the delivery is all the more important. Maybe you can remind them again of the importance of why they are here, but do not do this conspicuously. Do not draw attention to their inattention. Rapping on the table or clearing your throat with a, "Hmmm, Hmmm" is not appropriate. Saying, "Hello?" is rude and broadcasts rube.

MICROPHONES

When used correctly, microphones are wonderful. Given the option, I always prefer a microphone. If a microphone is offered, use it, even if the room would not seem to require it. The technicians who are responsible for the conference room know what is required. Even if you have strong vocal projection, use the microphone. It will ensure that those technical issues will be heard with the same clarity they are spoken. If a microphone is needed, check ahead of time to see if a wireless microphone is available or at least a microphone you can hold. If the microphone is to be held, hold it, and do not play with it or the cord. Microphone gymnastics are distracting and make unwanted sounds.

A Nobel laureate spoke at our Physics Colloquium. He is so famous that a natural phenomenon was named after him and its discovery. Reluctantly, I disclose neither his name nor the phenomena because of the respect I have for his accomplishments. Here is what happened. This wizard among wizards, scientist among scientists, giant among giants, lectured from the center of the stage holding a microphone. It was not wireless and the 50-foot extension cord coiled on the floor beside him and extended off stage-left. While he lectured, he periodically turned around and pointed to the large screen behind him, a screen that was simultaneously projecting his major points. He seemed able only to rotate in the clockwise direction. He would rotate around to look at the big screen and then rotate back around, in the same direction, to face us. Twenty viewgraphs into the presentation and the cord was coiled about him like a boa constrictor. He could no longer move as his feet were wrapped like Amenhotep II. Fortunately, a helpful graduate student freed him while we watched. Extricated and oblivious, he resumed the presentation as if nothing had happened. In another twenty viewgraphs, déjà vu. Fortunately, he was at the end of his presentation. As peers, we accepted his eccentricities—after all, he was a Nobel laureate—but such antics spell "failure to persuade" to decision makers. The Nobel laureate exemplified the stereo-typical wizard. His situation is a warning to you. Do not play with the microphone.

If the microphone is on a pole, remove the microphone from its pole and hold it. Move the pole to one side and begin. Use your free hand to ensure you do not trip on the cord as you move. If the presentation is less than five minutes, or if several speak-ers will be alternating, then (and only then) is it acceptable to just stand behind the microphone and leave it on the pole. To be continually taking a microphone in and out of its stand every few minutes is distracting.

If you must stand behind the microphone-on-a-pole, do not fig-leaf your hands in front or handcuff them behind. If you are not using them to gesticulate, then drop them at your sides. This feels awkward at first and you wonder where those arms have been all along. Practice until you become accustomed to having your hands comfort-ably relaxed at your sides.

The best solution is a wireless microphone, either lapel or headset. A lapel micro-phone should not be clipped at the throat, as it will amplify undesired noises like breathing and moving your head. It should not be clipped at the waist because it becomes directional, projecting loud as you glance down at your notes and becoming too faint as you look at the audience. This desultory change in volume is distracting.

The lapel microphone should be clipped near the center of the sternum with the barrel pointed upwards. This is one time when wearing a jacket is wonderful as the microphone can be clipped to it and the cord hidden underneath the jacket. Ladies should think ahead of time about what type of blouse to wear. Some transmitters are heavy and bulky or have a large cable connec-tion at the transmitter. This can clip on a belt, if you wear one, or clip to the seam of your pants or skirt. The weight is distract-ing, though, feeling like your pants or skirt need to be pulled up. If possible, try out the microphone ahead of time to get accustomed to the equipment. Be careful that it is turned on only when desired. There is usually an independent switch on the transmitter itself.

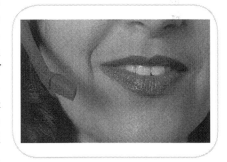

Use a microphone in larger conference rooms or when offered.

PODIUM

Never stand behind the podium unless the (only) microphone is attached to it and there is absolutely no way the audience reasonably can hear you, otherwise. Ministers and commencement speakers stand behind podiums. This is not the effect you desire.

Anything you place between yourself and the audience is a divider. You are on this side, they are on that side. There is a barrier between. You want to be on the side of the audience. If it is possible to do so unobtrusively, push the podium to one side. Otherwise, stand alongside it or in front of it. The best podium is the kind where a single pole supports a framed holder. It is light, unobtrusive, and can be put aside for easy reference to anything you place upon it. If you use one of these podiums, again, do not stand behind it.

I was at a conference with Texas governor Rick Perry. There were about five speakers before him. They all stood behind the large cabinet podium, leaned over, and spoke into the fixed microphone. Speaker number three even pushed the microphone aside, but remained behind the podium. This forced us to endure ringing harmonics for twenty minutes. When the governor spoke, he took the microphone in hand and moved away from the armoire podium. He stood in front, facing the audience, He moved, gesticulated, and did all the motions I elucidate here. The difference was astounding and the impact stunning. Move out to where the audience can see you.

WRONG: do not stand behind the podium. Stand in front or to one side.

HUMOR

If you say or do something really stupid and the audience laughs, laugh with them. If something untoward occurs and they laugh at your expense, laugh with them. I was giving a presentation to a government intelligence organization. I finished a half-hour review telling them how reliable and immune from failure our product was. At that precise point, the electricity went off in the entire building and stayed off about thirty seconds. This had nothing to do with our product, of course, but the audience burst out laughing at the humor. I did not find it humorous, but that was my mistake. I should have. It takes awhile to learn these things.

Sometimes it just is not your day. If so, laugh about it and just go on. There will be other opportunities. Do not take yourself so seriously. I was fifteen minutes into a presentation to the president (of our division) when the fire alarm sounded. We emptied the room and evacuated the building. Faulty equipment. We returned ten minutes later. I started again and five minutes afterwards the alarm sounds, again. We marched outside a second time. We were outside longer, so the meeting was just abandoned. I started the task of rescheduling. Sometimes it just is not your day. Learn to laugh about it.

If you know the audience, humor can be effective. Humorous anecdotes are certainly in order. Humor can backfire, though, if you do not know the audience.

Formal jokes are mostly tolerated, seldom effective. I recommend that you not use jokes.

With international audiences, forget jokes entirely. Smile, of course, and be pleasant, but do not experiment with jokes. Jim Smith, our vice president of marketing, was in Asia with our in-country agent. Jim was scheduled to make the marketing presentation. Jim is a big joker. He always has one in his pocket. Knowing this, the agent warns him, "Jim, it just will not be funny in Thailand. No matter how funny it might be to Americans, the humor is not the same. Do not tell one of your jokes."

Nevertheless, Jim is overcome (possibly encouraged by the after-dinner libations) and, during his presentation, starts one of his favorite jokes. In the US, it actually is very funny and fits well with his presentation. I hoped for the best. As our agent translated, the audience chuckled at the right time and laughed uproariously at the punch line, even clapped. Our vice president of marketing beamed with pride. I thought, "Well, so much for that."

Afterwards, I asked our agent about the joke. He said, "Oh, I didn't translate any of the joke. That would have been disaster. When Jim started, I translated, 'Mr. Smith is starting to tell an American joke and we need to show him what a good joke it is … now, Mr. Smith is getting to a funny part. Please laugh politely … now, Mr. Smith is finished telling his joke. Please laugh loudly and give him applause for such a fine joke.'"

NOTHING TO WEAR

If you are not dressed like the rest, you can feel awkward. You do not need this feeling to add to your problems. Wear clothes that fit. Baggy, sloppy clothes send out the signal that your work is sloppy. Do not wear styles or colors that draw attention to your person or attire. In order to persuade, you need to make the audience know that you are "one of them," not a wizard techie who just rocketed in from Uranus.

Dress modestly and professionally. Short skirts, plunging necklines, and tight-fitting clothing draw attention to yourself in an unbefitting way for a professional. Tennis shoes, jeans, tank tops, tee-shirts, and similar are almost always out of place. It was in Denver that I learned the dress code, "Wear a collared shirt," meaning not a tee-shirt. So, men, wear a "collared shirt."

If you are overdressed, it is straightforward to "dress down." It is difficult to reverse. Overdressed is never egregious and always easily forgiven. When in doubt, err on the side of being overdressed. Men can always take off a tie. Both men and women can remove a jacket. Inquire with the administrative assistant as to the dress code for the meeting and follow that. If it requires buying new clothes, well, that is not such a bad thing.

For men, "business casual" means a jacket, no tie, slacks (not jeans), and dress shoes (not canvas or lounge). For women, it means either a jacket with slacks or skirt (or a nice dress) and dress shoes. Most men and a few women scientists and engineers look conspicuous in business casual because they have no taste. Go to a reputable store and buy a jacket recommended by a knowledgeable clerk. Have it tailored to fit. Buy slacks that actually go with the jacket. Men, purchase a white or pastel shirt to go with those and a conservative tie. Throw away the camel and plaid sport coats, dad's tie, the shirt that is two collar sizes too large, and, yes, the loafers.

On the East Coast, most meetings are business casual or may even require a matching suit. The West Coast has a dress code two notches down from what would be expected in the rest of the country. Jeans and a jacket are commonplace. It is always best to check ahead of time.

Internationally, the standard is usually a notch higher than in the US, especially if you are the visitor. Ask before you leave so that you can take the right clothes. Even though some of my favorite attire was purchased outside the US, the time required to purchase it is not always available.

Make no fashion statements. Wear appropriate attire.

Jeff Bezos, Larry Page, and Bill Gates can wear whatever they like. So can you when you get to their level. Until then, follow the decision makers.

STANDING ROOM ONLY

The worst environment for a persuasive presentation is the large auditorium in which the speaker must stand behind a cabinet-podium talking into a fixed microphone. You are forced to stand in one spot, the lights are turned down, and all eyes are glued to a projection screen. You are not part of the audience; you are standing apart. It is difficult to get buy-in.

In some very small or crowded conference rooms, there is no place to stand, or if you do stand, you stand right against the table, looking and feeling awkward. The best way to present, sometimes the only way, is to remain seated. In some sense this would seem favorable because it is more intimate, but the ability to control the presentation is weakened. There may be no projection capability and even if there is one, it can be behind you, considerably apart, or at an obtuse angle.

If you must make the presentation while seated, you can request that each attendee follow along with a hardcopy you provide. This close-in, seated presentation takes some getting used to. It is limiting. Arm motions are difficult. You are limited mostly to leaving the elbows at the side and using the hands and fingers. Otherwise, you elbow a decision maker. You cannot move about to emphasize change in thought. You cannot point to key sections of the presentation since each attendee is leaning over his own copy. It is critically limiting.

LET THERE BE LIGHT

Invariably, you begin your presentation with the room bright and cheerful and some well-meaning individual then turns off the lights. If you use projection to show text or graphics, someone will assume they must dim the lights. At a time when you most want the audience alert, taking notes, hanging onto your every thought, down go the lights. You do not want an environment of slumber and transcendental meditation. You want them alert and awake. You want the lights on. You want to be open, available, and transparent.

Do not dim the lights. Dimming has three effects, all bad.

Effect One: It lulls each to sleep, psychologically telling them that it's time to nightie-night. If you must make your presentation shortly after lunch, it will be tantamount to giving everyone a pillow. They will not awaken until you are finished. This is the last thing you want after all your hard work.

Effect Two: It pushes each to focus *continuously* on the projection screen and not on you. You want the audience focused on you, what you say, your motions, your emphasis, your pointing, and not the visuals. In a lighted room, you can point them toward the visuals, but in a dimly lit room they will naturally gravitate to watching the projected material alone and this will work quickly to pull them away from listening and push them to reading.

Effect Three: The only seemingly positive value to dimming the lights is if the visuals are too dim to see with the normal room lights. It is not positive. The audience of decision makers wonders why you let this happen and why you are such a poor planner. Either brighten up the viewgraphs by using different colors or ensure ahead of time that the projection is sufficient for your material.

Leave the lights burning brightly. The audience will have hardcopy in front of them, anyway. If a few of the viewgraphs are dim, then live with it and make sure it does not happen again. If they are all dim, then fix the problem before you go into the meeting room.

In very large auditoriums or those cheap hotel conference rooms, the projection capabilities are usually deplorable. If you know you will be presenting with inadequate facilities, try to secure a brighter projector. If this is known ahead of time, rework the charts to make them more compatible with the inferior display capabilities. Eliminate poor charts, as required. Do everything you can to keep the lights burning brightly while you speak.

If you have no choice but to contend with a dim room, it will be to your disadvantage. Make certain it is worth the tradeoff. Do not get too attached to those viewgraphs. Maybe you can do the presentation without some or all of them. It is better for the audience to be awake and active than it is they recognize every detail of your projected viewgraphs.

NOTES

By the time you organize the presentation, write and rewrite, practice and edit, and go through a Pink or Red Team, you know the content. If you make viewgraphs, these have bullets or points to be made, so they are your notes. You do not need additional note cards. Under no circumstances, whatsoever, should you read the viewgraph bullets to the decision makers.

You should not memorize your presentation except in the most formal of presentations or where legal exactness is required. Except among the elite, a memorized or a read presentation comes across as just that. Improvisation takes the skill and knowledge of the presenter, puts it together dynamically, tailors it to the audience, and drives home the end results you want. Do not memorize what you plan to say, with the exception of the opening lines and the closing lines as discussed earlier.

It takes a while to get away from notes, to achieve this competent and confident level of delivery, but it is an achievable goal. Until you achieve this level, you may feel the need for additional notes to ensure that you make certain points. In such a case, annotate your hardcopy of the presentation to emphasize certain bullets or remind you where to add additional details. Your annotated hardcopy can then be placed to one side where it is not noticed. If, while practicing, you find yourself looking at your notes constantly or shuffling them often, then you should practice until you need to glance at them only infrequently.

If there are no viewgraphs or information being projected and you are just speaking, then the standard 3"x5" index cards are suitable. Do not be conspicuous with the notes.

HARDCOPY

Information that exists in digital form is called *softcopy*. Softcopy can reside on magnetic discs, electronic memory components, and the like. The softcopy term probably first came about when computer information was put on magnetic film or flexible disks called "floppies," but it has continued in the jargon since and become the standard terminology.

The tangible product of all those digits taken out of the computer and placed in human-identifiable form is called *hardcopy*. Printed files, faxes and similar are called *hardcopy*. The printed text of your presentation is the hardcopy.

Always have sufficient hardcopies of the presentation so that each member in the audience has a copy. If some were printed in color and some black-and-white, then ensure that the key individuals have the color copies. With the advent of low-cost duplication equipment and the expectations of a modern audience, I recommend all copies be in color, but that may be cost-prohibitive. Do, always, make color copies for the key decision makers. Again, though, color should be used only where it is effective.

Do you provide the hardcopy at the start of the presentation or after the presentation? You provide each a hardcopy at the start of the presentation. Why?

The advantage of distributing the hardcopy at the start of the meeting is that it is available for them to a.) take notes and b.) follow along without having to read the projected charts. In a large conference room, with complex graphics and inadequate lighting, not everyone can read the projected material, effortlessly. Other advantages are c.) to have an annotated copy for filing and reference, and d.) to know how long the presentation will be (and how near you are to being finished).

Why is it important that they know how long the presentation will be? It will ameliorate one of your problems. They are worried about getting to the next meeting. If you make the content too long for the time allotted (shame on you), it will reduce their questions, possibly permitting you to finish, but not to succeed.

I do *not* recommend giving your presentation and then handing out hardcopy at the end. Why?

I was contracting to have some rental property painted. The contractor measured it all, walked to his truck and came back in three minutes with a wonderfully printed estimate. It looked like a fashion ad. He held it in his hands, tempting me. Nothing doing, but we had to sit down, together, while he read it to me, page by page. For the next 15 minutes he elucidated the skill of his workers, the quality of his paint, the advantages of spraying (versus brushing), color, the competition, and the chemistry of drying paint. I just wanted the house painted. Finally, he got to the price, the only thing I really cared about. How did I feel about him not giving me a copy in the beginning to page through? Antagonistic.

I go to purchase a used car. Is this a pleasant experience? No. Why? Because the first thing I want to know is the real price I must pay. After I know the price, I can reason value, price being only one of the components of value. But it will be an hour or so of haggling before we finally get to the turnkey price. What are my feelings? Antagonistic.

Machiavelli, in his famous treatise to the Medici family, admonished that bad news should be given very quickly so as to shorten the time of pain and that good news should be trickled out slowly, so as to lengthen the period of benevolence. The problem the audience has, in a presentation where they do not have the hardcopy up front, is that they anticipate that the news is bad. There is a feeling that it is "too much," either money, personnel, time, or whatever. They sense that you are holding back because you know they are not going to like the recommendations. The audience is uncertain. It leaves an innate feeling of distrust and apprehension. You do not want that.

In addition, if the decision makers do not have hardcopy during the presentation, they will not take notes. They will lean back in their chairs and relax. This is not good. You want them alert and interested. Since they do not have a copy of the presentation, it does force them to follow your presentation, as you present it, and not look ahead. "Yes, of course," you say, "I do not give it to them ahead of time because I want them to follow what I am saying." It is the same reason some presenters project only one sentence or bullet at a time on every viewgraph. The strategy is that you do not want the decision makers to see your cards until you play them. This is especially desired, you think, if your last few pages, the recommendations and actions, are con- troversial. It is the salesperson approach.

However, if you hand them the hardcopy only as they leave the room, they will not read it, they will not have not taken notes on it, and they will have no feeling of ownership. The only persons who will have any use for the hardcopy will be the administrative assistants who will immediately send them to the warehouse where Indiana Jones shipped the Ark of the Covenant. All other copies get trashed.

How to gain the most advantage? Distribute your copies *immediately* prior to your presentation and then begin without delay. This gives them no time to idly flip through it. If your presentation is swift and lively, they will follow along, nicely. Inevitably, there will be those who flip ahead—maybe it helps them organize their thoughts. Regardless, ignore that and stay on task.

POINTERS

In a meeting, there are few things more distracting than watching a speaker paint the room with a laser pointer. Up and down it goes, back and forth, like a light saber. Now we have Lissajous figures, now it climbs up to the ceiling, falls down to the floor, flies out the window. "Oh, what was he saying? Who knows?" Up the wall, down the wall, across the floor. A pirouette here, a touché there.

Suppose we take the laser pointer away from this wizard and give him a stick pointer, probably one of those collapsible-antenna ones. He will jerk it in and out like a pump jack. He will thrust it about like a Musketeer's foil or be an orchestra conduc- tor on steroids.

"What about a long Lucite pointer?" you ask. I answer, "Do you know it is possible to flex a Lucite wand between your hands and produce standing waves? Do you know how great it works as a cane while you are giving the presentation? Do you know how much it bends like a pole-vault rod? Do you know how many of these antics a wizard can produce each minute?"

Or worse still, you could wiggle the computer-mouse like a NASCAR driver at Talladega. Go faster and faster to emphasize the good points? No.

Whether laser, wand, or computer, whether metal, acrylic, or electronic, what- ever the pointer, use it delicately and sparingly, if at all. At longer distances it is absolutely impossible to hold a laser-pointer still. Most presenters are slightly nervous anyway and the ever-so-tiny movement leverages to gigantic proportions. In the hands of the uncoordinated, the acrylic wand can be deadly. I have not witnessed a person run through, but I have seen several pointers bounce off walls, flop on the floor, and skitter across the conference table—all to the defocusing of the audience and discomfiture of the presenter.

If you simply must use a laser pointer, hold the laser-pointer against your side and use the wrist to point. Your body will help steady the pointer. Also, you will be less prone to flip it around all the time trying to point out every pixel in the graphic. Use it sparingly and turn it off after the logical point it made. Keep it on for no more than five to seven seconds at a time. Do not switch it on and off like a signal light. Never wiggle it around for emphasis.

If using a wand, do not carry it around while you make the presentation. Only the most disciplined can resist telescoping the pointer in and out or waving it about. Whatever the constitution of the pointer, let it lie on the table or easel until you need it and pick it up only when required. As soon as its use is complete, replace it. If it telescopes, leave it in the extended position for the entire presentation. Just lay it to one side. When in doubt, do not use the pointer.

The best pointer is yourself. Most conference rooms are not so large as to require a mechanical or electronic pointer. Walk directly to the projected material and point with your arm and finger. Why is this preferable? This puts you *into* the presentation. You are now a part of it. The audience never takes their eyes off of you, which is what you want. Get "into" the presentation. Now, you are focusing them on what you want, the presentation. This is extremely effective. It not only eliminates the bad habits invited by the pointer but keeps you, the presenter, in control. Many professional speech coaches agree with the concept of using yourself as the pointer.

Without fail, if you do this, some helpful attendee will hand you a laser or telescopic pointer that she keeps for just such an occasion. (Can you imagine?) Accept the offered pointer willingly and at the first chance, lay it on the table in plain sight. Leave it there and use your arm and body. They can see, then, that you have the pointer in plain view, but have chosen to not use it. The ignorant but Good Samaritan will think you just keep forgetting about it.

Using your body as the pointer will automatically limit how often you point and it will totally eliminate the Lissajous laser or challenging sword. Do not stand directly in front of the projected light or you will be spotlighted and block out any presented material. Back off to the side and point *into*

Point with your body. Get into the act.

the presentation. You do not have to put your finger on the exact location, just direct their eyes. If the room lights are still on, as they should be, this will not seem awkward and it will make the presentation flow smoother. As you come away from the chart, the audience will follow you again as you move to your next element in the presentation.

Creating Leverage

- Attitude counts.
- Be bold, but not arrogant or superior.
- Use body language to your advantage.
- Examine carefully the body language of the audience and adjust your style accordingly.
- Use a microphone when offered, preferably a wireless one.
- Jokes are bad. Humor is good.
- Eliminate notes whenever possible. Make them otherwise inconspicuous.
- Wear clothing suitable for the location and the level of the decision makers.
- If at all possible, stand when you present.
- Keep the lights bright. Never dim them.
- Pass out hardcopy of your presentation immediately before you begin.
- Whenever possible, do not use a pointer. Use yourself.

Chapter 12: External Factors
Creating the Environment for Success

The half minute that we daily devote to the winding-up of our watches is an exertion of labor almost insensible; yet, by the aid of a few wheels, its effect is spread over the whole twenty-four hours.
— Charles Babbage

All pendulum motion is described by a physical law relating the period of the pendulum to its length and to the local acceleration due to gravity. The amazing part is that the motion should be independent of the mass of the pendulum bob and the amplitude of the swing. How was this law discovered? A young Galileo Galilei is attending mass at the Duomo in Pisa. A large censer, an incense burner of Goliath proportions, hangs suspended from the nave of the arch, high above. About fifty feet to one side of the censer is an upper, narrow balcony. In every service, the priests enter onto the balcony and with a long cord pull the censer over to where they can light it. They light all the lamps and the incense, releasing the burning chandelier to swing back and forth, back and forth, far above the heads of the congregation. The sight of the swinging censer with its dark, smoky clouds, the sound of the sizzling embers, and the odor that fills the nave stays forever in the mind. I know. I was awed when I saw it for myself and envisioned the pensive Galileo sitting right beside me. Maybe Galileo would have formulated his theory in any setting but he did not. The setting was crucial for him and so it will be for you.

The most frustrating task for a technologist is a task that does not involve technology. To the wizard, "non-technical" equates to "not important." Unfortunately, if we are to persuade decision makers, we must delete that entry out of the *Wizard Dictionary*. Technology persuasion requires some non-technical tasks and some non-technical skills. If success is your goal, and what else is there, then there are non-technical tasks and techniques that must be an integral part of your game. They are essential.

SETTING UP THE MEETING

It would seem obvious that if you are requested to make a presentation, someone would contact you with the information about the meeting. Correct? Not necessarily. You cannot assume so. From the very beginning of the assignment, find out *why* the decision makers want the presentation, *who* will be at the meeting, *when* the presentation is expected, and *how much time* you will be given to present. Often, the assignment of a presentation comes about as the action of some other meeting. If you are in that meeting, ask questions while everyone is in the room and make sure you know the *Why* of the requested meeting. The very worst situation and, unfortunately, all too common, is to be given the assignment by a superior who took the action and you receive only a general idea of what is expected. Go find out. Make sure you know *exactly* what they want to hear.

If the presentation is to a group that meets periodically, like a board or review team, then these meetings are scheduled formally and it becomes a matter of getting on the agenda for the next meeting. In such a case, contact the administrative assistant for that group. If the scheduled meeting is too soon for you to prepare adequately, request postponement to the subsequent scheduled meeting to give you sufficient time to prepare. Enumerate and quantify reasons why this would be more advantageous to the *attendees*. Senior decision makers are usually indifferent to the inconvenience it may cause you, so stick to the items of interest to them—better data, test results completed, or similar, but specify with numbers how much better, what tests, when, and the like. This may or may not result in postponement, but give them the opportunity to make the decision.

Often, you are requested to make a presentation and it is your responsibility to schedule the presentation with the administrative assistant of one of the executives. In such an event, procrastination is deadly. Be certain to work the schedule, immediately, within the hour, if possible. Why? The available time of executives is very limited. If several of them must meet together to hear your presentation, then scheduling is complex. If you wait even a few days to schedule the meeting, you may find that the next available time slot is too far in the future. So, you say, "That is great. It gives me more time to work on the data." Wrong. The problem is, your opinion does not count. Why did the decision makers ask for the meeting? Is there an investment opportunity that is fleeting? Are there competing options being evaluated (last in, first out)? Are there decision makers who are required for a decision, but cannot make a later date? Delay can be deadly to the very thing you want to promote. You have more time to work on the presentation, true, but to what end? Circumstances overtake you.

In this case, *you* have a problem, not the decision makers. The logic that "if the executives want to hear it, they will make time," is naïve. Schedule it immediately.

As much as possible, make it part of your job description to know the administrative assistants. They are invaluable personnel, take their jobs very seriously, and will be a great help to you in preparation. They are often some of the nicest people in the company. And, if not, work around that and get it done. Request a time slot that fits the window given to you. If such a slot is not available, inquire if any of the decision makers can reschedule to accommodate the assignment given to you. Often, it can take several days just to get a meeting "locked in." Do not underestimate the power of the administrative assistants to move meetings around if you are in good stead with them. And, do not underestimate the difficulty of getting a meeting with their boss if you come across as arrogant and disrespectful of their position.

Anticipate that the original meeting date will change, especially if you do not want it to change. Every executive has unplanned demands. A technology presentation is seldom number one on the list. Remain flexible. If the meeting date is changed, usually your presentation is moved to a later date, but not necessarily. So, you should plan your own internal schedule to have your presentation completed a day or two prior to the scheduled meeting. Even if the meeting date does not change, you can use this extra time for polishing, maybe even have a second Red Team.

The administrative assistant will ask, "Who should be at the meeting?" Hence, you will need to know who the executives intended to invite. Do not say, "I don't know." It is not the administrator's job, it is yours. Go find out. All of this is difficult and awkward for a technology wizard, but it is critical and essential. I am fully aware that the typical wizard thinks this is beneath him, or her, and should be done by someone else. Nevertheless, if you want your recommendations followed, take charge of details.

If at all possible, have the administrative assistant of the key decision maker send out the meeting invitation to the other decision makers. The other decision makers and attendees will be much more responsive, and considerably more accommodating, than if the meeting invitation comes from a subordinate.

LAY OF THE LAND

As soon as possible, find out the number of attendees and the *Reconnoiter the lay of the land.*
location, that is, which meeting room. What equipment is available? Find out. Do not assume that every meeting room will have high-speed internet connections, high resolution projectors, remote controls, microphones, and lighting controls. Most conference and meeting rooms were designed several years ago and retrofitted as technology advanced. If you are outside the US, then you must ask specific questions about power outlets and available equipment, personnel, everything. If you plan to battle, you need to know what the battlefield looks like.

Knowing the "lay of the land" is essential to success.

When Pompey's army moved against Caesar at Pharsalus, Pompey had over twice as many soldiers and horsemen as Caesar. Pompey, however, had removed himself hundreds of miles from his naval supply chain, whereas Caesar's men were well-conditioned to fight on meager rations. This battle marked the turning point for Caesar and the demise of Pompey because Caesar took advantage of the lay of the land.

At Marathon, the Athenian general, Miltiades, was outnumbered 30 to 1 by the Persian king, Darius. The Persian empire controlled a territory nearly seventy times as great as did the loosely confederated Greek city-states. The hoplites attacked on the open plain of Marathon because they knew the lay of the land. They knew that archers and cavalry, of which the Persians had many and the Greeks had none, would be virtually ineffective against the shock warfare of highly-disciplined armored men in close phalanx formation.

And what if you ignore the lay of the land? Over 350,000 men were killed at Gallipoli, the almost impregnable entrance to the Dardanelles. Knowing the lay of the land is critical in war. It is also crucial to making a persuasive presentation.

The best solution to an ill-equipped room is to bring all the equipment yourself, often just a laptop and a projector. Mechanically, this is not difficult given the small size of laptop computers and projectors, but operationally this may not be permitted for security or protocol reasons, or it may just be impractical for other reasons. Blank walls make marginally acceptable projection screens. Regardless, find out the limitations and know these ahead of time so you can fashion your presentation accordingly.

Executives generally have well-designated, specially equipped meeting rooms. This is not always true of subordinates. Therefore, when you brief the subordinates be sure and find out what equipment they have. You may be surprised at how frugal many well-known companies, and even the government, can be. Most think of the Pentagon as an icon of the latest and greatest in techno-wizardry. Prepare for a shock at the warren of cubby holes used for offices and meeting rooms. If you know ahead of time and can prepare for it, all obstacles can be overcome. But, not knowing can be fatal. Uncover the lay of the land before you stumble onto the terrain.

Senior executives often have full-time personnel employed just to run meetings. Imagine. In this case, find out who that person is and talk with the individual. How early do you need to show up? Is there opportunity for you to practice in the room, maybe early in the morning or the night before? What type of media is required to transfer your presentation to the projection equipment—memory sticks, disks, and similar?

If you travel, it is often difficult to bring sufficient copies. Can copies be made, internally, at the company site and long before the meeting? Or, perhaps you need to go to an all-night copy service. Fortunately, at least one major copy and collation company is open 24/7 and has offices in most major cities. If you write a few large proposals, you will know the pleasure of working there at 2:00 A.M. to complete material for delivery at 8:00 A.M.

VIDEO AND AUDIO CONFERENCING

It is commonplace to have individuals at the meeting who are either on the phone or videoconferencing. This produces additional problems. If the key decision maker is on the video, this is particularly difficult. Ensure that you know well ahead of time who will be at the meeting location itself and who will be on video or teleconference.

In most videoconferences, the video-attendee will have several video "windows" on their computer, showing either the meeting room, the presenter, or the current viewgraph. If the viewgraph is to be readable on the video window, that usually eliminates other simultaneous windows. If the attendee is unimportant to the decision, that is not so critical. If the attendee is an important decision maker, then it is paramount that the words on the charts give all the essentials. Make certain that the video-attendee has a hardcopy of your presentation *prior* to the meeting. If you wait until the meeting starts (forbid) and then try to send it (electronically, of course) this will take valuable time from your presentation and the transmission is plagued with last minute "gotchas." Plus, the video-attendee must have a copy made while you start the presentation. It is awkward, unprofessional, and defeating. The typical wizard just views this lost time as techie-at-work, but decision makers see it as stereotype. The video-attendees need a hardcopy in front of them so they can pay more attention to the other video windows and not see just viewgraphs. They need to become part of the presentation. Strange as it may seem to a technology wizard, all of this is your responsibility—at least to ensure that it is done. Do not settle for, "I told someone to do that." It does not happen that way in the real world. Follow up and make sure everything is in order *before* you enter the conference room. You will need to follow up as many times as it takes to get it done.

We are on vacation. It is early morning in Seattle. I am called to a teleconference with the president and his vice presidents, being notified a few days prior. The family is with me in the hotel room, getting dressed, waiting for me to finish the teleconference so we can parasail Puget Sound. No video is available, just audio. I am the presenter and all the other attendees, every one of the decision makers, are in the president's conference room in Dallas. It is the worst of situations. I had sent all the attendees an electronic version of my presentation, at least. The administrative assistants had made hardcopy available for everyone (which I had ensured, in advance). Nevertheless, trying to effect persuasion from a telephone is massively stressing. You have very few tools with which to work. You cannot read their body language. They cannot read yours. You cannot keep them "on the same page." You cannot point to key items. You cannot work through complicated graphs, or tables, or charts. You cannot look them in the eye. The phone is a weak duplex, mostly handling only one-way communication at a time. You have trouble interpreting the gist of their questions. It is easy for them to just "color in the square" and say they met with you. Their inclination is to postpone a decision, a negative action for you. Needless to say, it is frustrating and disappointing.

If at all possible, meet in person with the *key* decision makers. Persuading the key decision maker by videoconference or teleconference is limiting and, if the decision is heavy, reduces your chances for success. Few recommendations are ever so compelling that they are won simply by having someone read the viewgraphs, look at the charts, and hear the explanations. These presentations are not a complete report. Much of the information comes from the spoken word, the answering of questions, and, especially, the spontaneous interaction. Persuasion involves so much more than bullets and viewgraphs. Remember that.

CONTINGENCIES—PLAN FOR THE INEVITABLE

For those who have never traveled to India, there is the impression that the entire country is third-world equipped. That is not a correct viewpoint of every locale, especially Bangalore. It is astonishing the growth in the number of of technology companies located within that region, fast becoming the Silicon Valley of India. They specialize in telecommunications, software, and outsourcing.

Nevertheless, while there, I was called to make an impromptu presentation to the mayor. Bangalore is a massive city so this is roughly equivalent to making a presentation to the mayor of Los Angeles. In India, however, the relationship between civilian and military is different. This particular person also was head of certain government intelligence activities for several major cities in India. He was a key decision maker for our technology. A half hour prior to the meeting, we were at his *classified* office (not his normal civil office) setting up equipment for the presentation. We were allowed into the room, under guard, only 30 minutes before the meeting started. Because of security reasons, there was no wireless internet connection that could be made available to me. I was not permitted to use my own computer for similar reasons. This was not a major factor, however, because I had the presentation on a memory stick. This turned out, in spite of the homework and coordination I had done ahead of time, to be incompatible with their computer system. (You can get wrong information.) We were unable to get their classified computer to read either that memory stick or the two backup disks I had with me (their special security software). I had brought my own high-resolution projector but, now, no way to show the data, and they were concerned that the projector had memory or recording capability (it did not). There also was no screen in the room, anyway, as their practice was to project onto the dingy, dimpled wall behind me. Time was now a waning commodity. Disaster was looming. The final solution was to have them drag out an antiquated view-graph machine and show overhead transparencies. Unbelievably, I had brought along a set of transparencies for just such an emergency—joking to the team that we were now in the six-sigma region of risk avoidance. Six-sigma or not, it was required. It was not pretty, but it worked and shows how with even the best of planning you need to plan for contingencies.

Actually, I had yet another backup should that plan have failed. I carried spiral-bound copies of the presentation and would have given each of them a copy to follow while I used an easel. The small size of the room would have made that acceptable and it was the hardcopy I planned to leave, anyway. Lesson to be learned—prepare for the worst of contingencies. Prepare for them, because they invariably will happen. The more important the presentation, the more likely their occurrence. We succeeded because we had backup after backup. We planned for the unexpected, so there was no unexpected, just undesired. Equipment does fail, usually at the most inopportune times.

High quality projectors are amazingly small, relatively inexpensive, and phenomenally durable. I find it easy to carry my own projector and laptop, just in case. But things happen. Bulbs burn out. Did you bring a spare? Disks and memory sticks can have problems. Did you bring redundancy? The laptop can have problems. Is there a replacement? Plan for all contingencies and expect them to occur. You will not be disappointed.

LAY THE FOUNDATION

If your recommendations are either broad or deep, ensure that you brief the decision maker's staff well in advance of the main meeting. Not persuading the staffers and underlings can

> *Secure the underpinnings. Brief the staffers and subordinates of the decision makers. Brief any other technical experts who either factor into a favorable decision or represent strong contention.*

ensure defeat with the decision maker. It may determine whether or not you even get to see the big boss. This is especially true in government, where you cannot approach the official without a recommendation from the staff. The bureaucracy mechanisms vary from country to country, but every agency has bureaucracy. Briefing of the subordinates has a complexity of its own.

I met with Saudi Arabia's military intelligence. The key decision maker was the head of military intelligence, another royal prince (of which there are many). The position is similar to the US Joint Chief of Staff. I was to brief him on the technical details of a commercial satellite we were launching. Our in-country agents spent weeks setting up the various meetings. We briefed all the major staff leads before we had the meeting with the prince. This occupied a week of preliminary meetings, receptions, and dinners.

When finally we were ushered into the conference room with the prince, we found it filled with the subordinates we had briefed. The staff was divided almost equally in thirds, one-third favoring our proposal, one-third objecting to our proposal, one-third neutral, and three-thirds silent during the entire presentation. Any decision had deep and broad implications for Saudi Arabia.

Although the staff was mute, it was absolutely essential that when the prince looked around the room, all the subordinates should nod to indicate their organizations had been briefed and had an opinion. I suspect (as in many analogous decisions in the US) that only one opinion mattered, but those other briefings opened the door to the prince and perhaps his "ears to hear." At a minimum, it "colored in the squares" that were a requirement, there and everywhere. The subordinate organizations must be briefed ahead of time. On rare occasions, one can cut through the red tape and get a direct audience with the decision maker. Strangely, this is not a guaranteed success as I will relate.

Through a series of events, our team briefed, "privately," a US multi-star general on a new counter-intelligence capability. He was a key commander and the meeting was set up through a congressman. We had opportunity to brief only a very few of the general's underlings prior to meeting with the general himself. He loved the capability we presented. He loved it. He was excited, which, for a senior military officer, is only a tenth on the Richter-excitement scale. He looked at me and said, "Let's do it." "Fantastic!" I said. We went back to our desks and waited for the check to arrive. Months later, we were still waiting. It never came. Why?

First of all, the general was not an engineer, so he may not have understood all the intricacies. Second, he may have assumed all of his team was in concurrence when, in reality, only some of the team had been briefed. (Never mind what we said. What did he hear?) As is often the case, there is much infighting within the ranks, not so much because there are two sides to every argument, but because there are two sides to every door, an inside and an outside. We were on the outside. As soon as he left with the driver, the factions roiled. Those not briefed began to show ways they felt our capability could be achieved internally (internal to the command), with no need to hire an outside contractor like ourselves. Technically, this was not true and history proved us correct, but being right is not a measure of success.

Had we been given opportunity to brief his entire staff ahead of time, we could have had a reasonable chance of pointing out weaknesses in their arguments, both technical and political. It is probable we could have won them over or teamed with them. As it was, their constant access to the decision maker and close infighting trumped reality. The lesson to be learned—brief the subordinates, thoroughly. To the extent possible, incorporate their modifications. Above all, get their buy-in. If they are not in agreement, then begin to work compromises or supportive arguments. Know the *political* lay of the land, as well as the physical lay of the land, before you brief the key decision maker.

Briefings to congresspersons often require that the staff members, themselves, rewrite the proposal and take it to the congressperson in their own words, with their own spin. This, of course, is how it is done with the very top decision makers, yet it serves to illustrate how critical it is to brief the staffers. Do not go into a meeting with senior decision makers in any organization, unless you have briefed the subordinates prior. Achieve consensus, effect a compromise, or, at a minimum, differentiate the areas of disagreement. All the underlings need not agree with your proposition. However, you need to resolve as many differences as possible before the main meeting and, above all, know the pressure points *a priori*. Usually, you have one chance only with the senior person. Make it count. The worst mistake of all to have the staffers come to the meeting and tell the boss, "This is the first I've heard about this." That is tantamount either to a death knell for your project, or a long road to repentance.

The very best outcome, of course, is to have the staff fully back your presentation and take part as if it were their own. Occasionally, this happens and, if so, you can celebrate your good fortune. I would not stock champagne, though. It is a rare occurrence. As your presentations grow in technical complexity, as your implementation requirements increase in magnitude, the less you can expect unanimity. Take success in strides. Most successes come only after hard work.

Finishing Up

- Ensure that you schedule the meeting immediately.
- Know *Why* you are being asked to present.
- Know who should attend and ensure they are invited.
- Know the lay of the land, the conference room for the meeting, and all the support equipment.
- Prepare for any videoconferencing or teleconferencing.
- Contingencies are the norm. Prepare for them.
- Brief the subordinates to the key decision maker and get their buy-in.
- At the very least, understand the pressure points of conflict with the subordinates. To the extent possible, resolve them prior to the meeting.

Chapter 13: The Elephant Dance
Sidestepping Confrontation and Conflict

Tact is the art of making a point without making an enemy.
— Michael Faraday

*E*arlier, we touched upon confrontation. It is such a critical element of persuasion as to require a more detailed study.

Decision makers who are interested and contemplative ask questions. If your recommendations have value, they invoke inspection. Queries indicate the decision makers are assessing your recommendations. They may desire clarification, validation, or reassurance. They may want to ensure they understand all the ramifications. No request of importance will be accepted, without probing.

The worst situation is for the audience to be mute throughout. No presentation can be so comprehensive as to intrinsically answer all questions. Silence is a dirge that mourns a dead project—yours. You need questions. Learn to covet inquiry.

But, how to answer those inquiries? Except in the most formal of occasions, you should reserve and take time for questions, either during the presentation, afterwards, or both. The latter is optimum. If there are no questions at the end, that is a bad sign. It can mean one of four things, either 1.) you failed to make them understand, 2.) you talked so long and gave so many details that they just want to leave, 3.) they came in with their minds made up and you could not have persuaded them, anyway, or 4.) they are occupied with other thoughts from which you were unable to remove them. None of these scenarios is favorable.

One way to prevent an embarrassing silence at the end is to draw your audience into the presentation and not wait until the end. When you feel you have made a salient point, you might ask the key decision maker how she feels about what you just said. Does she agree? Are there points that need refinement? It is always good form to ask the audience, "Is this understandable?" They will usually respond in some way as to indicate which direction you should go.

You can pose your own question and then answer it, like Hamlet does in his soliloquy. When employed *sparingly*, this technique is effective by adding variety to the style of the delivery. However, it does not sufficiently invite interlocution.

If the audience is not responsive, do not resort to crass interjections like, "Hello???" Or, "Are you still with me?" This puts the audience in opposition. Do not challenge the audience, draw them in. Do not ask them questions for which they might be embarrassed.

In the rare instances when the decision makers agree to your recommendations without inquiry, you can gather your chips, cash them in, and get out of Vegas—but silence usually does not herald agreement. When necessary, engage the decision makers. "Mr. CEO, did I answer your question fully?" "Is what I am presenting what you expected to hear?" "Are you surprised at these results?" "Were you expecting lower manufacturing costs, higher performance, or better results in the phase one trials?" "Do you feel we have sufficiently justified the cost?" "Is anything lacking that you need in order to make a decision?" You will need to read body language. It is not likely you will get a direct answer to these questions, but that is not the purpose of asking. The purpose is to draw them out and find out where they stand, so you can begin to fashion your arguments to their needs and convince them of your recommendations.

If one of the decision makers begins inquiry, others will engage, often conversing with each other—in which case a good presenter must also be a good listener and an efficient moderator. Know how to move the discussion and interaction into your next point and do not let these discussions take on a life of their own. Bring out the salient points, but never lose control of the meeting or ignore the clock. Just because the president is arguing with the CFO does not mean you will get more time for your presentation. Do not interrupt decision makers. Wait for an opportunity. Learn to work the audience. You may have to be quick. Sometimes, you must take control, but do this professionally and with gentle persuasion.

A room full of decision makers is a room full of egos, business politics, abstruse personalities, and combative fiefdoms. It is a study in ultra high-energy interactions. The inter-dynamic comments of the executives will do more to influence persuasion than anything you say. Use their comments to modify your words, *in situ*, to continually tailor your presentation to their specific needs.

With such a collection of Type A personalities, disagreement may erupt between them, perhaps leveled at what you are presenting, perhaps not. It is best to let the giants fight among themselves and not join the fracas. The war games are too unknown to you and you are too insignificant a combatant. Your participative comments will neither be welcomed, appreciated, nor apropos. Stick to your presentation and do not take sides in the conflict. Do not engage in inter-fiefdom peacemaking. Learn to differentiate between a war of particulars (something you presented) and the ongoing war of kingdoms.

I was in Riyadh, Saudi Arabia presenting to their equivalent of our US National Geospatial-Intelligence Agency. Our in-country agent briefed us, prior, on the inter-relationships of the participants. It was a weak substitute for first-hand knowledge. In the US, with my many years experience in this field, I would have had considerable insight into the players and politics, but that was of little value there and even could have been misleading. The presentation was in English. All their side conversations and comments to each other were in Arabic. It would have been inappropriate and rude to get a translation each time, because if they had wanted to converse in English for my benefit, they would have done so. According to the agent, the meeting went well. Such feedback often is self-serving, though, so it is hard to know.

The bottom line is this. Do not wander among elephants. Your puny little technical bones will be splintered.

Be honest and forthright, but circumspect. When decision makers argue among themselves, hold tight and wait until they give you the floor again. Often, their argument has little to do with what you said directly, and everything to do with the private agendas of the players in the room. Shrapnel can fly everywhere, so keep a low profile until you get the nod to go ahead or until there is sufficient lull to attempt a restart.

WHEN AND HOW TO ANSWER A QUESTION

When a question is asked, it should be answered immediately regardless of where you are in the presentation. Otherwise, the impact is lost. It is weak to table a question and say, "I'll cover that later." You can do that, of course, but the moment will be lost and the audience interest turned tepid. The significance of the answer will be forgotten. Seize the moment and make your points when they are best made, by succinctly answering the question when it is asked. When you do perhaps cover more details later, say simply, "These are the details to the question we discussed a few moments ago." Then, summarize briefly the answer and go on. Repeat things only as necessary to keep the flow integral and smooth.

In a discussion between you and several in the audience, or between the audience members themselves, listen to all sides and be objective. If the discussion leads you to change your mind, then do so. Do not stubbornly hold to opinions or positions. By the same token, do not be fickle. If you have a position and the data back it up, stick to it. Remember, data and evidence notwithstanding, the decision team may make a decision contrary to your recommendations. They must, after all, run a business.

Strive to make the audience interaction integrate with the presentation and not degenerate to unnecessary elaboration. Make your answer cogent and compact. Bruce is a hardware wizard in my organization. The vice president came up to me one day and said, "If Bruce were a piano salesman and I asked him, 'What piano do you recommend?' he would start by telling me how trees grow. He would continue with the Bessemer process, the stress upon strings, and the current world supply of ivory substitutes. All I really wanted was to buy a piano for my daughter." This is an example of turning a simple question into a research topic. The tendency to provide excess detail is the number one criticism of technologists, engineers, and scientists, so be sensitive to your leanings.

Often, someone in the audience will ask a question that was answered, already. The inquirer may have been asleep, or maybe she did not associate the prior answer with her question. Your response? Just answer the question. No need to say, "That's the answer I gave earlier," implying, "You should have been listening (dummy)." Be polite, and answer the question asked without belittling the inquirer. Your answer may not have been as clear as you surmised.

Speakers use the cliché, "There is no such a thing as a dumb question." Well, of course there is. And the dumbest of questions may come from the chief decision maker. If so, this is usually a sign that you are doing a poor job of communicating. Regardless, answer the question correctly and tactfully, and think about how you could better articulate. If the decision maker thinks it worth asking, answer it. Maybe your presentation needs more work.

Do not make put-down statements like, "That's the same question asked a few minutes ago. I covered that while you were out of the room. After what I presented, how could you think that?" Again, simply answer the question. (And, yes. These all were flesh and blood, real-life wizard comments I recorded.)

A skilled wizard will lead the audience through each viewgraph so that a question naturally arising on one viewgraph is answered on the next viewgraph. In other words, one viewgraph leads into another. Often, someone in the audience will verbally ask the very question that you are about to answer. Avoid gleeful and trite remarks like, "You're a good straight man. I answer that in the next viewgraph." The proper way to handle this is to begin the answer and bring up the next chart so that it all continues to flow smoothly and rewards the questioner for a well-thought-out question.

Do not refer to a question as smart, good, or dumb, no matter what its merit. It is problematic to judge, "That's a good question." Saying this one is good implies the former one was bad. If you, the presenter, think it such a good question, why did you not think of it yourself and answer it already in the presentation? I understand the intention is to interact with the audience, but decision makers do not require additional ego reinforcement. Most presenters make it repetitious and wind up telling everyone, "That's a good question." "That's a good question." Avoid all this by not grading the questions.

Sometimes if you stop and answer the question, it will absolutely sidetrack the entire presentation. In such case, it usually is permissible to inform the questioner that you will answer that in a few minutes. If you do this, though, you must remember the question and when you get to that point later in the presentation, specifically mention that it answers the question asked earlier by such-and-such. If the person who asked the question is a key decision maker, then it is good to ask the questioner if your response sufficiently answered the question. Do this sparingly. Do not open Pandora's box.

WHEN YOU DO NOT KNOW THE ANSWER

What is the best response when you do not know the answer to a question? A professor of mine once admonished, "If you don't know the answer, just keep saying right things and they will give up thinking *they* haven't communicated." I do not like that approach as it puts the blame on the person asking rather than you, the person who has not the answer. Politicians ignore questions entirely and begin answering a question they wish you had asked so they can give you information they wish you to know. Another expert recommended saying, "I don't know, but I will find out and get back with you." This beats nothing, but I do not agree with any of these.

The first two are unacceptable because you are the wizard and you owe the inquirer a valid answer, especially if you expect success in the meeting. The problem with the last one is that there is no mechanism for effectively inserting the answer back into the decision process. A decision deferred is a decision lost. By the time you find the answer the next hour or the next day, the audience will have forgotten the question or its relevance. The answer, even if apropos, often brings up a whole new set of questions that must go unanswered. As a member of such an audience, I have received answers by email within hours of the actual presentation, but the moment was lost. Being a decision maker in the meeting, I was off onto other things. A solid answer during the presentation would have been salient. Afterwards, it was flat. If I was not the one asking the question, I may have little interest, or forgotten even why the question was important.

In a technology presentation, the person presenting the material is necessarily foisted as the expert. So you need to study and prepare and be able to answer any reasonable question. Then, invite one or two members of your own team to attend the meeting with you. They should bring detailed information with them and be looking for answers to complement your responses as the presenter, if additional information is required. A good team will know when to answer for the presenter. You can look at them or call upon them. This helps build the team as it gets other technology personnel in front of management and also makes for a solid front. It lets another person speak and tends to keep the presentation fresh and active, the audience alert and responsive.

Do everything you can to prepare so that you know the answer to any question that might reasonably be asked.

Having said all that, if the question is relevant and neither you nor your team knows the answer, then respond, "I do not know, but I will find out and send to (be specific as to whom) an email (be specific as to what mechanism), by close of business tomorrow (be specific as to when)." Then, respond as promised.

Thus, state precisely who will receive the answer. Is it only the person asking the question, the management team, all the audience, or whom? Specify how you will respond. Tell them that it will be an email, a report, another meeting, whichever best satisfies the intent of the question. It is of little value to answer to persons not interested who will have forgotten the question, anyway. Set the time for the response. Make it as soon as you can, but in no case a shorter time than is required to get the right answer, get it in the right form, and verify it. You do not want to hurry a bogus answer that you also have to correct. That approach will minimize trust.

If the answer is broad or deep, you may want to request another presentation that elaborates on that answer. If you can answer it in due course, well and good; if not, search for alternative ways. Always remember that the purpose is to get agreement to your recommendations. Do whatever it takes.

If you receive more than two relevant questions that neither you nor your team can answer adequately, then you either did not prepare sufficiently, or you prepared for the wrong subject or for the wrong audience. Do not make excuses. Use this as a learning tool to be more prepared the next time. Not every presentation will be dynamite. Everyone, even the best, fails to be adequately prepared at some time or another. Make this a rarity. View all failures as positive experiences and a chance to improve your skills and eliminate mistakes.

At the end of a good presentation, during the final question and answer period, it is easy to fall into a flippant or silly attitude thinking all danger is past and you can celebrate. Do not get out of character. Remain diligent in all the rules we discussed, as if it were still the presentation (which it is).

MODERATION

Wizards can be obnoxiously impatient with those who do not follow their explanations. It is common for them to address confrontation with superiority. Perhaps it is subconscious. Think-tanks are full of posters and cartoon clippings that illustrate. I look at my organization. One office has a 2'x4' poster of a huge gorilla. The caption reads, "If I want your opinion, I'll beat it out of you." Another is simply a warning, "Fools entering are not entertained lightly." Another is a Dilbert cartoon that goes, roughly, "When will scientists stop wasting their time with solar power and start working on the only unlimited supply of energy on the planet: stupidity."

If a wizard takes this attitude into the conference room, as too often occurs, it catalyzes confrontation. Solon, the lawgiver of Athens, admonished, μηδεν αγαν, *nothing in excess*. Think of this Greek's wisdom before you answer. Your response should be circumspect, anticipative, polite, and professional. Your presentation style should be dynamic and energized, but be careful such that in answering questions your energy is not channeled into affront.

When answering a question, remember what we said earlier, "words matter." Defuse a confrontation by selecting non-confrontational, courteous, and respectful words. Be as matter-of-fact as possible in your rendering. Remember, even among scientists, very knowledgeable people can look at the same data and come to different conclusions. Allow your audience that same privilege. Take care not to be judgmental or superior in tone or intent. This is difficult for a wizard, so work on it.

NO DIRECT CONFRONTATION

Solid information will ignite new technology concepts and ideas. A byproduct of this combustion is controversy. It is rare to present data so compelling and so overwhelming as to quell every naysayer. If you are a technologist, scientist, engineer, mathematician, or similar, you cannot avoid confrontation. But you can avoid war.

My friend, Bill Daughton, would say, "A child of four would know I'm right." In other words, "It's a no-brainer." Well, maybe, but maybe not. Regardless, dilute dogmatic assessments, as managers and leaders can be unbelievably opaque toward what technologists would consider transparent. It is not as obvious as you think. Technology can be complex and the business into which it melds is complex. Confrontation will arise. How does one handle direct confrontation?

Sir Liddell Hart was a lifetime military strategist who wrote numerous books on strategic encounters and war. While military editor of the *Encyclopedia Britannica,* he began a broad and comprehensive survey of military tactics used in every major conflict since 490 B.C. He concluded that in the hundreds of major conflicts he studied, less than half a dozen were won by direct confrontation. With few exceptions, major battles are won, he concluded, not by direct confrontation, but by *indirect maneuvers.*

Sun Tzu, in his 500 B.C. treatise, *The Art of War,* noted, "The highest form of generalship is to balk the enemy's plans; the next best is to prevent the junction of the enemy's forces; the next in order is to attack the enemy's army in the field; the worst policy of all is to besiege walled cities. In all fighting, the direct method may be used for joining battle, but *indirect methods will be needed in order to secure victory.*"

The great Byzantine strategist, Belisarius, had this to say about confrontation, "The most complete and happy victory is this; to compel one's enemy to give up his purpose, while suffering no harm oneself."

In a technology presentation, one compels victory by giving succinct answers based on substantive facts and apropos logic.

I drove our rental car from Istanbul, along the Edimine Canakkale Yoli highway to the ruins of ancient Troy. Leaving Istanbul, the road winds down the Gallipoli peninsula to where the divided continent necks together at Kalitbahir. From there, a ferry diesels cargo across the narrowed Dardanelles to Canakkale. Our journey occurred, coincidentally, one day prior to the commemorative anniversary of the WWI Gallipoli campaign. The buses we passed were full of young Aussie and British sailors who had slept on ship and were making their way to the memorial site.

The Gallipoli campaign was a nine-month battle that cost 130,000 lives and wounded another 260,000. The Allies coveted the Dardanelles to capture the Ottoman stronghold and ship supplies to Russia. When you see this peninsula, you will be as perplexed as I was that a military strategist would order such a direct attack upon so impenetrable a position. As the Allies began their attack, German Admiral de Robeck exclaimed, "Gallant fellows, these soldiers; they always go for the thickest place in the fence." It was immeasurably costly, this direct frontal attack. And so it will be for you, if you try to resolve technology confrontation by direct attack.

Anticipating confrontation reiterates the requirement to have meetings, Red Teams, and to brief the staffers, before the main meeting. Make every effort to defuse, compromise, or democratize the confrontation *before* the meeting. Make every effort *in* the meeting to avoid a direct frontal attack. Use your energies to work out compromises, at least during the meeting itself. Inveigh the opponent *a priori* to permit you to give your side of the story, without interruption, and then offer to give him the same courtesy should he decide to contest your results. If these are high-level decision makers and you reviewed your presentation prior with their staff, then you could start the decision maker meeting by stating that you received the support of their teams in preparing this presentation or that their teams disagreed, whichever. Be candid.

When you know that confrontation in the decision maker meeting is inevitable, begin the presentation by seeking common ground. Do not start anywhere near the controversial components. Begin with the common elements upon which all agree. If the company has been successful with product X and you are proposing improvements that might be controversial, then by all means start out with how successful product X has been. Start with common ground and attempt to garner support as you go.

This "common ground" is like the green zone in certain armed confrontations. Both sides respect the green zone under established protocols. So, start your presentation in the green zone and establish some facts, or principles, that all agree on before you introduce controversial issues. Let them know you are one of the good guys. You are on the same team with them, which is different than trying to get them on the same team with you.

As early in the presentation as possible, get favorable audience interaction with the things you both agree on, you and the decision makers. Getting them to support even part of your presentation is a strong start. Attorneys do this as a matter of training. When the prosecution and the defense begin jury selection, the *voir dire* proceedings, the defense will finish with something like, "And, if the facts show that my client was not intoxicated, then what will your verdict be?" You must answer the question, and the only logical response is "Not guilty." Whether the defendant is guilty or not, the defense attorney already is getting you to agree, "Not guilty." You want to do a similar thing regarding controversy. You want the decision makers to agree with you on something You want them to feel comfortable agreeing with you so that when you hit the controversial portion, they are attuned to being on your side.

Do not start the presentation by separating the sheep from the goats. Do not bring up anything that will push them to take an early stance. Doing so will cause any opposition to dig in their heels and every statement you make, every piece of data you present, from that point on will drive them in deeper. Work hard to keep decision makers in the Green Zone. Human nature is so strong that changing a mind, once set, requires multiple times the energy of just setting it. Think if you attached a chain to your car and tried, by hand, to pull it to the end of the driveway. If the driveway is downhill, climb aboard. You do not need the chain. On level ground it is difficult. On uphill terrain, even the world's strongest man will sweat and strain.

But, let an occupant of that auto apply the brakes and you cannot budge it, even if pulling downhill. Do not give opportunity for the decision makers to anchor a position against you.

Do not wander into unknown territory as this will cause the decision makers to discount your already controversial arguments. Stick to the subject and the things you know well and are prepared to discuss.

Stay nimble but sure-footed. Listen to their objections. Do they have merit? Are they correct? Should you modify your thinking? Do not make the mistake of digging in *your* heels and being unmovable. Maybe a compromise is in order or a total change of mind. Maybe now is not the right time for anything. Maybe here is not the right place for a decision. Maybe you are not as prepared as you thought. If so, admit where you are and use that to get them on board with you. What can you do to strengthen your position? What is lacking that would change their minds? See if there is a path circumventing the obstacles.

If, upon careful examination, you and the decision makers are at a communications impasse, then, at least, try to come out of the meeting with some measure of success. In a football game, the offense may spend five minutes getting to the opponent's 20-yard line and find they are fourth down with eight yards to go. Fourth down go-for-it's are low-percentage football. Rather than come away empty-handed, the team decides to kick a field goal and at least get a few points on the board. Do the same with controversy. Even if the decision makers cannot agree to your recommendations, try to come away with points on the board.

Close decisions very often are decided by feelings, impressions, and history. Also, remember that trust factor. At the very least, leave the meeting with trust. Do not let the emotions of the moment cloud your vision that there will be other opportunities.

Sidestepping Confrontation and Conflict

- Conflict is inevitable. Try to resolve the major impediments before the meeting.

- When at all possible, answer questions when they are asked. Do not defer them.

- If you do not know the answer and must delay the response, be specific as to how, when, in what manner, and to whom you will respond. Make sure you follow through, immediately.

- Keep the answers professional and matter-of-fact. Never belittle or try to intimidate the audience. This will backfire.

- Get on common ground as quickly as possible with the audience.

- Do not engage in a strategy of direct frontal attack.

- Put some points on the board. Make every meeting successful to some extent.

- Avoid emotional responses.

Chapter 14: Postmortem
Learning from Errors

A person who never made a mistake never tried anything new.
— Albert Einstein

Wizards participate in wizard-conventions. These conventions, or societies as they are called, are organized around different technologies. This particular one, The Electrochemical Society, focuses on the physics and manufacturing of semiconductor devices. Our presentation is scheduled for twenty minutes. My associate and I "tag-team." That approach alone is a gutsy idea for this group. Fifteen minutes into it and we are just another presentation to print up in their minutes. Then, we ended with a 90-second video, not of our successes, but of all the things that went wrong, machines breaking, gooey "resist" flying everywhere, and silicon wafers smashing into a thousand fragments all over the laboratory, complete with the sound effects. It is the only *scientific* presentation, ever, for which I will receive a standing ovation. Yet, what brings them to their feet is not our profound discoveries, but all the things that went awry. They identify with failure.

All performers love the spotlight and the applause. They crave the thrill of victory, and the gold-medal ceremony. Champions train, sweat, torture, and persevere—to win. Everyone loves to win. Yet, for every winner, there must, by definition, be a loser. Sometimes, you lose.

No one likes to lose but wizards agonize for weeks over the most trivial of mistakes. On an outer level, wizards feel they are right (read 'perfect'), in a broad sense, and take it intentionally and intensely personally when their recommendations are not approved.

Strangely, they often do not recognize whether they have even won or lost. For example, the decision makers can approve nine out of ten items but that one tiny thing they did not approve can drive the wizard to complete distraction. Or a wizard may go the other way. Nothing is approved, but because the decision makers are supportive, complimentary, and sympathetic, the wizard interpretation is a "successful meeting." There is no understanding the irony that a wizard can envision victory as defeat or defeat as victory.

Most often, though, no matter the actual outcome, the feeing is one of having lost. Sometimes you do lose, big time, for real. Do not dwell on failure. Yet, neither should you ignore it as if it never happened. We must examine present failure to ensure future success.

I encountered the term "postmortem" in middle school. We published a weekly newspaper. I was a reporter. Our advisor, the venerable Mrs. Mildred Huguelet, was a peering, exacting, and detailed commander. We published our newspaper on Thursday afternoon. On Friday afternoon, we sat in our little news factory, a converted classroom lined with inks and papers, and examined the shortcomings of our latest issue. She called this meeting our "postmortem." She felt there were always things to improve, things to correct, things that went wrong. She said we needed to understand what we did wrong as well as what we did right. What we did wrong might even be more important than what we did right. It took me a long time to really accept that.

Immediately after a presentation, one can feel various emotions—elation, mortification, or anything in-between. Sometimes, all these feelings are concomitant; some things went well and some things did not; some were successes, and some disasters. A day or two after the presentation, but not the same day, sit down with a few peers and perform the postmortem. Do this for both triumph and disaster. Follow Rudyard Kipling's advice to "treat those two imposters just the same." It is better to perform the postmortem with a team, particularly if they were in the meeting. The team can help put things into perspective that an individual may distort. If possible, include a senior leader on the postmortem team.

Review the list from the postmortem and apply it in your next presentation. Then, have a postmortem on that presentation and so on. Do this every time, throughout your entire career. Be honest and objective in every instance. Not too harsh, not too gentle. You can never drive the list of shortcomings to zero. There is always room for improvement. As you increase in knowledge, capability, and position within the organization, your standards and expectations will rise. As your vision grows, so also the hurdles, so also your ability to scale those hurdles and attain the goals. This is why I said before that you cannot take things personally. You cannot get down on yourself. You will be forever improving, constantly getting better at communication. As long as you are doing something, you will make mistakes. So what?

MAKE A LIST

Start with a list of what went right. Even the worst of presentations has some good points. The list may be short:

a.) The meeting ended (finally).

b.) I'm still alive.

Avery Johnson, a longtime coach in the NBA and sports commentator, admonished his team, "Don't get down on yourself. When you make a bad shot or turn over the ball, whatever, just go on to the next shot. Don't get down on yourself." This is sage advice from an expert. You can help no one, including yourself, by moping about. One coral does not make a reef. Start with what went right.

Perhaps the *Why* segment was extremely good or the audience liked the recommendations in the *How* segment. Perhaps the graphics were good or a Visual *Aide* was particularly effective. Maybe you did a credible job on the *What* portion. Maybe it was just the right length, the audience was responsive, the audience asked the right questions. Make a list of what went right and reinforce those things in your next presentation.

Make a list of what things went wrong and why. Use your frustration and disappointments from the meeting to forge new capabilities. Do not waste energy on blame, excuses, complaints, anger, frustration, or despondency. Become a professional. Put *all* of that energy into improvement.

In asking "why" you made mistakes, you will need to go down four and five layers before you get to the "core" reason. It goes like this. For example, you note:

"The decision makers did not accept my recommendations."

Why? (the first layer)

"They did not believe that we could save that much money with my new idea."

Why did they not believe it? (the second layer)

"I ran out of time before I could explain all the details."

Why did you run out of time? (the third layer)

"I tried to show them a bunch of numbers and should have put them into a financial spreadsheet."

Why did you not put them into a financial spreadsheet? (the fourth layer)

"I did not know exactly how to do that or exactly what they wanted to see."

Why? (fifth layer)

"Because I did not want to ask for help."

Now you have gotten to the core reason that your recommendation was not accepted. Had you stopped at the first "why" you would have concluded that either the decision makers were daft or you simply were not persuasive. The real problem was that you did not put the data in a form they understood. And, you did not do that because you did not want to ask for help. You need to spend the time to get to these core reasons. Otherwise, you work on the wrong things. When you list what went wrong in the meeting, keep asking "why" until you get to the core reason.

TOO MUCH DETAIL

What went wrong? The most common thing for wizards is that you really did not have time to finish and had no real plan *for* a finish. There was too much peripheral detail. Perhaps there were simply too many viewgraphs. One of the reasons for this excess is that technologists seldom get to make presentations. Thus, the tendency is to dump the entire load and make up for all those times when technology was shortchanged. It does not work that way, unfortunately. Estimate the percentage of excess. Often, this can be as great as 50%. In other words, your presentation was 50% longer than it should have been for the time allotted. Read Chapter Seven on allocation and planning. Next time, shorten the presentation. Above all, do not be fooled into thinking that the same approach will get different results next time. Shorten it by the amount it was too long.

Shortening a presentation is not easy because you must present sufficient data to be convincing. Then, how can you be convincing in the time allotted? Usually, one cannot usually just omit things. "I'll show this, I'll not show that," usually does not work because the result is disjointed and unconvincing. The data must be compressed, summarized, coalesced, and simplified. This is where you, as the technology wizard, can outshine your counterparts. The ability to distill detailed technical jargon, data, and concepts into simple, understandable language is invaluable. Refer back to Chapter Eight and our discussion of how to take the best advantage of Visual *Aides*, viewgraphs, graphics, and words.

Again, do not think that the next time you will just *do* better. Go to the next meeting with fewer charts, much fewer. If those prove too few, then when you analyze that meeting maybe you can afford to add some for the next presentation. Having too few is easier to repair than having too many. The message is, where there are shortcomings or errors, *do something* about it. Do not expect that it will resuscitate spontaneously just because you wish it *so*.

The postmortem feedback may tell you that you were too nervous, forgot key points, or just did not present it adequately. Much of this comes with practice, so how would you fix that? Practice. Do not *wish* you were better. Do something to be better. Read the sections in this book and then apply them. Practice at home. Practice in the shower. Maybe you should give the presentation to a Pink Team and ask them to critique your style. Put yourself forward to interact and gain the confidence you need. The message, again, is to do *something* about it.

DEVELOP THE WILL TO SUCCEED

Examine the other areas that went wrong and think how you could have changed them to be successful. Do not lull yourself into thinking, "No one could have persuaded this group of decision makers." You failed … this time. Admit it, and find ways to succeed next time. It can be done even with this group of decision makers. Go do it.

Most technologists hate this corrective feedback part. They innately believe that the facts and the data sell themselves. Think about it, though. Such logic often is not valid, even among scientists and engineers. It certainly is not true with decision makers. Plus, decision makers have different criteria and different thought processes. Go back again and read Chapter Two about the thought processes and personalities of decision makers.

Your ideas are worth selling, so sell them. If your ideas are not worth the effort, why not get another job and do something that is worth the effort? I know you want to go back into the laboratory and just get to work with more technology, but if you do that, the world will be the less for it. You must work through your own inhibitions and disappointments. You must go back and try again, with more ammunition, with different ammunition, or with a different weapon. Nothing of value comes easily. Quit whining and get to work on persuasion.

LOOK FOR COMPROMISES

What were their real objections? Work on those. If possible, meet privately with the key decision maker or equivalent naysayer and find out what you need to change in order to be successful. Was it that you asked for too much money? If so, examine your cost estimates. Get help from someone in finance and determine if you have been realistic in your assessments. If the costs were, indeed, realistic, then what can you cut out entirely, or where can you compromise? Can you make the existing equipment work another year and postpone that expense? If you hired the "right" people, could you get by with three instead of four? Can you work two shifts instead of one? Three, instead of two? Overlapping shifts? Shared equipment? Shared personnel? Can you do the job with cheaper or less experienced personnel? Maybe the job can be done by a technician instead of an engineer. Carefully consider all your alternatives.

Are there competing projects that can be combined with yours and attain synergy? Can some other source of funding be applied for part of the work? Can some of it be outsourced cheaper?

Most wizard technologists want to say, "It cannot be done any other way. The window of opportunity is closing. We either do it now, this way (meaning *my* way), or it cannot be done." Nonsense. Look for solutions that are sellable, ones that will get agreement from the decision makers.

OUT OF SYNCHRONIZATION

Maybe the decision makers do not have the money at this time. Then, find out when they will have it. Investments such as capital equipment, or research and development have an annual cycle and there is a schedule for that cycle. Find out the process and get in synchronization with it.

Money comes and goes. Funding opportunities come and go. Since a rising tide lifts all boats, whenever possible, time your presentation to take advantage of this. Look for the crests in funding and jump on a big wave and ride it to success.

It is 11:59 P.M. in Truro, Nova Scotia, at the northern inlet to the Bay of Fundy, on a chilly October night. Our group huddles in the cattails of a narrow tributary to the bay. A half-moon sheds some light and we have our flashlights. We are whispering and quiet as church mice. I hear a gentle rushing. Softly. It swells louder.

Faster, louder, it is approaching and picking up velocity like a torrent. It now sounds like the roar of a giant waterfall. Here is what is happening. The moon's gravity is pulling on the ocean water, pulling that water all the way into the Bay of Fundy, pulling all that water to the very top of the Bay, and funneling all that water into our narrow inlet. We shine our flashlights upstream and there it is—a two-foot-high tidal bore rushing upstream towards us, a wall of water, like something out of *The Ten Commandments*, but smaller. It is a tiny tsunami rolling along. We scramble up the bank and watch the bore's profile. The wall moves past us at eight miles per hour as we silhouette it. It is not a wave. It is a ledge of water two feet higher on one side than on the other. The moving vertical partition is the bore. Within five minutes, a lake has formed and water is roiling up the banks. We head for higher ground so we are not swept away.

Funding can be like that: predictable, yet transitory. You sit with a project and have no funding for a long time. Then, it seems like opportunities just come rushing by. You must be prepared. Get in synchronization with the timing cycles with the decision cycles, be it funding, capital equipment, personnel, R&D funds, or decisions of any type.

CONTRAST AGAINST COMPETING PROJECTS

Maybe the decision makers do not have the money at all. Why not? Are they spending it on competing projects? Can you make your argument so compelling that they drop funding for those other projects to support yours? This is done all the time in business. It is called winning and losing, cutting the losses.

Weigh your project against competing ones. Put arguments together in the language of the decision makers. This is a key time to ask for ideas from the staffers, those reporting to the decision makers. It does little good to come into a meeting, though, simply petitioning suggestions. Do your homework and put together some ideas. Then be prepared to modify them with other inputs.

IMPROVE YOUR RETURN

Perhaps your project does not show sufficient return on investment, or the time period is far too long. What can you do about that? Can you accelerate the research? Can you get a smaller return, but sooner? Can you do less, make less, but get to profitability faster? All of these things are considered by decision makers. Create reasons for them to reconsider your project. It is like a court case that suddenly finds new evidence. The new evidence must be considered. Find new evidence that addresses their objections and then petition for your day in court.

Analyzing What Went Wrong

- You learn more from one mistake than from ten successes.

- Never give up on a good idea.

- Develop the will, endurance, and perseverance it takes to be a winner.

- Make sure you do not kill your decision with peripheral details.

- Look for compromises that maintain the key elements of your recommendations.

- If you are requesting funds, ensure that you are in synchronization with the regular funding cycles.

- Investigate competing projects to see if your idea can supplant those.

- Find ways to enhance the return to the decision makers.

Chapter 15: Nightmares
Solving Special Problems

I have had dreams and I have had nightmares. I overcame my nightmares because of my dreams.

— Jonas Salk, MD, discoverer of the first polio vaccine

*I*t was a dark and stormy night. Incoherent. Bad things. Ghosts and ghouls and all types of indescribable fears mixed up together. Our noble wizard awakes in a sweat, mind racing, hands trembling, blood gone from the extremities. What was it? It returns in full force. Ah, yes. A nightmare, again. This time, there was not enough data, yet all those decision makers were chanting, "You must present *now*! You must present *now*!"

The night before, it was a different nightmare. She was on the platform and said something entirely wrong, but did not realize it until the meeting had ended and all the decision makers were out the door.

The night before that, it was worse. She was halfway into the presentation and suddenly saw a major error in her logic and, at that epiphanic moment, believed her entire logic to be faulty and incorrect.

And then, there was last week. The decision makers disagreed with her conclusions, even as she kept screaming, "Look at the data! Just look at the data! Anyone could see it if they would look at the data!"

Or the one when …

Such are the nightmares of the wizard. Special and unique nightmares, common to our race, innate in technology experts. This chapter goes through some special nightmares and posits solid solutions, solutions that start in the mind and attitude of the wizard presenter.

This book is written *for* wizards and *about* wizards. Wizards have special talents and special gifts. They also have special needs. Every wizard knows the unique problems that belie our race. Every wizard in engineering, science, and technology knows the fears that develop into nightmares. We address some of those problems here. While we cannot cover every horror, it is anticipated that the solutions given will provide insight to solve others not discussed.

LACK OF SUFFICIENT DATA

What should you do when the data are not sufficient, not the right data, or the data are otherwise inadequate, but you must make a presentation anyway?

First of all, no wizard ever thinks there is enough data. The engineering mentality is to never be finished. There are always improvements that can be made, modifications to consider, enhancements to fashion. We discussed that earlier. Always, there also are decision makers who want an answer and want it *now*! The question here is what if the data really are inadequate to make conclusions or recommendations. What if they are, in fact, so poor as to prevent you from even assigning a probability to the outcome?

For a start, be realistic. Engineers, scientists, and technologists do not want to commit themselves to conclusions until the data are one-hundred percent complete, everything is analyzed, all contingencies investigated, everything reconciled. Make sure this is not what is happening. Be honest and realistic in your assessment of the situation. Confer with peers, but be absolutely certain that their comments are not simply placating or sympathetic. When you look at the data does a competent set of peers agree that the data are inadequate to make the conclusions requested by the decision makers? If so, what do you do?

Start by making a short, succinct paragraph, or a set of bullet points that describe where you are in the data process. Define, exactly, what you *can*, at this point, conclude about the data. Do not focus on what you do not know, or cannot do. What do you know and what can you do?

Then, write a second paragraph, or a set of bullet points, describing exactly what is being requested by the decision makers. *Why* are they requesting the meeting? Be exact and precise.

Now, compare the two. Where do they differ? Then compose a third paragraph, or set of bullet points, describing exactly why the *decision makers* should consider postponing the meeting. Estimate when you will be prepared to provide the requested information. This is not a time to say, "I don't know, it's research." They will request someone who does know. You are the technology expert. When do you *think* you will be ready with the information they request? Make your paragraph factual, succinct, and candid. When you have completed your paragraph, look and see how it compares to the *Why* of the meeting, your second paragraph, above.

It is common for a wizard to put down all the reasons for postponing the meeting only to discover that the reasons do not even align with the purpose of the meeting. It is making excuses, not giving reasons. Check again for alignment with the purpose of the meeting. Does your paragraph address the purpose of the meeting, as stated? If so, get this statement to the decision makers and see if they agree that the meeting should be postponed to the date you recommend.

Rescheduling is not always feasible for a couple of reasons. First, the decision makers may be irate and now want to hear exactly why you are not ready. If so, they have changed the *Why* of the meeting. Now, you can, indeed, answer the question, because you know exactly why you are not ready. Second, you may have only one opportunity with the key decision maker and this is it. You can explain why you are not ready, but it may be to no avail.

Do not establish the reputation of never being ready, never having enough data, never being able to make a decision. You are the expert. You are the professional. Do your job.

Modern companies are driven by multiple competing factors, quarterly and annual reports, the competition, the market, the finances, the shareholders, unions, contracts, legalities, and so on. Technology is somewhere on the list. Today's speed of business will never allow the time for one-hundred percent confidence. That is why the decision maker is so important to the company and why they want to hear your presentation at the time scheduled. It is time to make a decision—the company needs your technical help to do so. Help them.

The best way to treat honesty is to quantify your results. For example, you might say, "Our goal was to double the capability of our product. To date, we have implemented only eight of the twelve changes required. Based on our analysis, we believe we can achieve our goal; yet, at this time, because of limited data, we have only a 60% confidence. We believe we can achieve 95% confidence, but it will require three more months of analysis and will cost an additional, unplanned, $250,000." This type of quantification gets the decision makers what they need to make a decision. They may not like it, but you have reported it accurately and honestly. Quantify your honesty. Do not cloak honesty with other issues. Do not allow optimism to overly favor your results. Do not permit pessimism to pad them. Make the best assessment you can and quantify it. A decision maker can work with quantified honesty, especially if you have developed that sense of trust we discussed.

Identify what is lacking, but be sure to quantify what will be required to correct the deficiencies. More time, larger staff, increased budget. Again, no padding and no cutting to the bone, either. I realize that some managers expect "padded" numbers, so they cut in half whatever you give them. I recommend you establish a reputation from the beginning, of being realistic and fair: no padding and no pretensions. Wizards are not very good at playing these money games, so do not play. Stick to your expertise and let the decision makers stick to theirs. If the budget is cut in half, then focus on what you can do with half the budget, not what you cannot do.

AN ERROR CREEPS IN

What do you do if you say something entirely wrong in the presentation only to realize it later?

Those who are not wizards have no concept of how demoralizing even the smallest error can be to a wizard. Wizards build their entire lives around an exact and correct science. They expect perfection from themselves. Errors that would seem unimportant to someone else are devastating to a wizard, absolutely mortifying. Wizard mistakes are nightmares.

Let us assume, first, that *during the presentation*, itself, you realize that you said something incorrect, earlier in the presentation. If the error is trivial, absolutely a nit, and of no real consequence, just go on. Do not waste their time and energy on minutiae. The problem with the wizard mentality, of course, is that no error seems trivial or unimportant. Wizards have tremendous difficulty distinguishing degrees of error. Molehills are mountains and mountains are molehills.

If the error is of magnitude, it needs to be corrected immediately. Stop and correct the error. Tell the audience that you just realized an error. You made a mistake. Here's what I said, here's what I meant to say. No need to launch a diatribe of *why* you said it. Everyone makes mistakes. Correct it and go on. You misspoke. Do not let the decision makers leave the room with incorrect information. You want them to have solid facts before they make a decision and you want to be the person who provides those solid facts. You want to develop trust.

Now, however, what do you do if you realize *after* the meeting that you made a serious error *in* the meeting? The first step is to do nothing. Yes, nothing. Do not go try to find the decision makers in the hallway on their way out the meeting and try to correct it. *Their* minds are off on other things and they are not in a receiving mode. *Your* mind is not in an effective transmitting mode, either. First, before you do anything, stop and think. You just finished a very trying event, at least for the technology wizard. Rest a moment. Take a breath. Check with your peers. Go to the office or lab and recheck the data. Ensure that you did, indeed, make a mistake. In the nervousness and tension of the presentation, it quite often is the case that you *think,* afterwards, that you made a mistake during the presentation when, in fact, you did not. This is, oh, so common. Check and recheck to make sure. You do not want to make two errors. If you discover that you did make a significant mistake in the presentation, *quantify* the error.

Write down exactly what you said and what you should have said. Read what you have just written. Have some knowledgeable peers read it. The wizard tendency is to overheat and write something more confusing than what was said in the meeting. The wizard reaction is to flood the page with details that decision makers will not understand. Let someone with a clear head read what you have written. Most often, they will need to reword it so that it speaks to the decision makers. You will probably need to explain why you misspoke and what caused the error.

Then, take what you have written, get with your management and see what actions should be taken. The selected course is usually to get the information to the decision makers quickly, possibly in an email. If at all possible, all of this needs to occur within an hour or two of the meeting. You do not want incorrect information to be further assimilated. You do not want the decision makers to make decision-errors or change policy based on the misinformation you gave. Correct it quickly, but quantify the correction in solid, decision maker language.

Understand why you made the error and put together an action plan to prevent your making such an error in the future. Perhaps you should have had peers in the conference room to help you. Are you are too much of a Lone Ranger? Maybe you were too biased, had too heavy a stake in the outcome. Maybe you were just nervous and misspoke. Maybe you just made a mistake. Whatever the reason, put corrective action in place. Do not quit trying. Learn from your mistakes. Everyone who does anything makes mistakes.

What do you do when the error is not just a mistake? What do you do if, after the meeting, maybe a few hours, or a few days, you realize that data you gave, or conclusions you made, were entirely wrong, and will critically and adversely impact the results?

This *is* a nightmare. First, do not panic, react, or take any action untowardly. Do yourself no harm. Do not write your letter of resignation.

There is an exemplum that applies here. You have heard it but you need to remember it at this critical point in your mistake.

An Eastern monarch asked his sages to compile the wisdom of all the ages. Expecting volumes, the monarch waited for months. They returned with a single sentence, "And, this too shall pass."

And, your situation, also, will pass. In some cases, the consequences can be devastating, but they will pass. This is not a time to panic, not a time to act in haste or out of emotion. Write down, exactly, what you said in the meeting, the data and the conclusions you gave. Write down what is different, now, and what new information you possess. More than ever, this is the time to quantify the error. Make sure it is quantified: how much and when? Have knowledgeable peers review it. The wizard tendency is to overcompensate and to make things worse than they are. Use sound judgment. You do not want to be labeled as one who overreacts. Reword the paragraph until it is succinct and exact. You will probably need to explain *why* you did what you did.

Then, take what you have written and get with your management to see what actions should be taken. The actions will be different depending upon the magnitude and the impact of the errors.

Understand why you made the error and put together an action plan to prevent your making such an error in the future. And, remember, this too shall pass.

What do you do when you are partway into your presentation and you suddenly question your entire results, everything you did?

This happens all the time to wizards. I understand it happens to performers, also. You are right in the middle of the presentation and someone asks a question that triggers the reaction. Suddenly, you detect, or think you detect, cracks in your data and faults in your logic. You point to a chart, observe something you never noticed before and, lo and behold, you question your entire results. In other words, right in the middle of the presentation, what do you do if you think you may have discovered a critical error? Strangely, this feeling is not unusual. It happens more often than you think. You may have been intimately involved in the data collection, but were buried in the trees. Now that you are presenting and you look at it from a distance, you begin to see the entire forest and notice something that went undetected heretofore. What do you do?

First, do nothing in the meeting. Continue on as if no epiphany ever occurred, because … it may not have. There may be no epiphany. It could be an illusion. The reason you do nothing, at this point, is because you do not know, for certain, there is a problem. You spent hours working with the team on this project. Do you really think you found a fatal flaw during the last minute or two? You might investigate it later only to find out your presentation was not in error and was correct after all. In the heat of the moment, in your nervousness, you just *thought* you saw an error. A speaker in front of an audience, any audience, is under pressure. A wizard in front of decision makers is boiling in a pressure cooker. Do not make spontaneous changes or assertions under this pressure. Do not change what took hours to produce, just because you have an apparent epiphany. This is not the time to impulsively demur.

After the meeting, alone or with your team, rethink what happened and rethink your epiphany. If there truly are errors and issues, work them as we have described above. If there really is an error, before you announce it, you need a plan to quan-tify, mitigate, or correct the error. Be sure you have a resolution plan. More often than not, however, you think you have found an error and it is not an error at all. If you react too hastily, you look incompetent and not trustworthy. That is why, in the meeting, you do not launch off in a contrary direction as the result of instantaneous, seat-of-the-pants, half-baked revelations. In my career, I have had this happen three or four times. It was frightening, but none of these resulted in finding any errors. They were just illusions of the moment. Do not get me wrong. I have had many errors, but I never discovered even one of them when I was actually delivering the presentation.

What do you do when the decision makers disagree with your conclusions and recommendations?

You have done your best presenting your case, but the decision makers fail to enact your recommendations. This happens, of course. No one can win all the time. Be circumspect. What is the reason for their disagreement? Did they understand what you said? It is common for the technologist to think "The decision makers did not understand," when, in fact, they did. But, suppose they actually did not understand?

What did the decision makers say? I have been in technology presentations and, afterwards, asked the despondent technologist, "What did the decision makers say?" I find, usually, that the wizard does not know. She heard what she *thought* the decision makers said, but not what they actually said. Listen carefully. What did the decision makers really say? What reason did they give for not confirming the recommendations?

Do they need more data? Do they need more convincing data? Did they ask for specific things? Was it a question of costing too much? Do they have the money? Was the timing bad? Did they offer recommendations? It can be a single issue or a host, but start with what they actually said. It is unusual for everything to be unacceptable. After all, you did give a presentation for some reason. So, what did the decision makers like and what did they not like?

What did they *not* say? Was there body language or unspoken words that conveyed another message? (Be careful here. It is common for wizards to think they are disliked. Focus on the content, not feelings or personality.) Was your presentation on the same page as the decision makers? Were you wasting their time? Were they just listening so they could "fill in the squares" and say they reviewed your presentation? Was the key decision maker in the meeting or just underlings with no authority to commit? These are difficult questions to answer. You must answer them mostly by body language and by doing your homework on the decision makers.

I have found, almost invariably, that decision makers *do* understand the presentation, but decide against it for other reasons. Make certain you know their reasons for rejection, whether they say them in words or otherwise. Is it repairable? Do you have the time, resources, will, and capability to make the changes they would require? The worst thing to do is correct what you *think* they wanted only to go back and discover you are still rejected. Before you make further investments, made certain you understand exactly why you were rejected and exactly what, if anything, is required for agreement.

If you are involved in research, you may not get answers to these questions. You may find that the decision makers simply reject your proposal and recommendations. They are simply not ready, for a multitude of reasons. The decision makers may fully understand your presentation, yet lack the wherewithal to state what is needed for them to agree. They may only know that they are not willing to follow your recommendations as presented. They may not have the vision for technology that you do. They may not see what is on the horizon. They may consider that your job. They may miss the opportunity or they may go with someone else. Whatever the reason, you must learn to make valid assessments. Whatever you do, maintain your value to the organization. Seek compromises wherever reasonable.

Oh, Those Nightmares

- Quantify. It is much easier to work with quantifiable recommendations than vague premises.
- Do not compound errors. Make certain they are errors.
- Correct erroneous information in a timely manner.
- Maintain integrity. Continue the relationship of trust.
- Understand why your recommendations are rejected.
- Evaluate verbal and nonverbal reasons.
- Carefully assess what it will take to get agreement.
- Learn to compromise.

Chapter 16: Wizard Success
Breathing at the Top

Don't be afraid to give up the good to go for the great.
— John Rockefeller

Y ou implement the techniques. You hone the skills. You achieve success. You establish a reputation as a trusted technologist and an effective communicator. You are asked, repeatedly, to make presentations. You are promoted to greater responsibilities. Your new jobs entail supervising other technologists. You receive promotions and a title. Which title does not matter. You are responsible for broader components, more comprehensive projects, larger funding, and greater risk. You answer to more powerful decision makers. Their numbers are greater and their questions more frequent. You keep climbing. What now?

To begin with, always remember how you got to where you are. It was not alone. It was with the help of technologists and decision makers. They took you in, as partner. They overlooked your mistakes and focused on your successes. They seasoned the good and mitigated the bad. Replicate their actions in yours.

With increasing authority comes increasing responsibility. All technology problems rise to the top. Everything that goes wrong is now your responsibility, maybe even your "fault," whether you even knew about it or not.

A good friend of mine was a senior officer in the company. He supervised thousands of personnel in a massive organization around the world, much of it in technology. An employee committed illegal acts throughout a period of many years. That individual is in prison. My friend is without his job. Why lose his job when the perpetrator had other managers, was four layers down in the organization, and the two had met only once? Because, it was judged my friend's responsibility "to know." And, too, someone had to pay the piper. Responsibility ascends to the apex.

You make a successful career as a technologist. You are at the top. All those personnel issues that you hate now land on *your* desk—reviews, hiring, firing, calming the seas, putting oil on the water, water on the fires. Those tasks once considered insignificant and menial now occupy more and more time and become the "important" tasks. You are shoved farther and farther away from the real technology. Now, you can only talk about what *others* have done. In the broad sense, they did it under your direction, but for a technologist, this is cold comfort. You do not get to do the "real" work anymore. You get to *hear* about the real work. You get to *assess* the real work.

You look around one day and realize you have become—the decision maker. You decide technology. You decide what is brought forward, you decide what is rejected. You decide what to promote and what to diminish. You fund this one, and extirpate that one. You have budgets and schedules to meet. You have suppliers and contractors to satisfy. You worry about intellectual property, about partnerships, and about teaming agreements. You critique the competition, give opinions, make judgments. Whether you like it or not, you have moved over to the side of the decision makers. You now know their side, at least in part. You have not switched, exactly, but the line of demarcation is not as distinct as once it was.

You may not be involved in technology details, but you are making decisions at the top. You are, in some sense, steering the ship. It may be a little ship, but it is going somewhere and it is going there under your direction, under your command. You are compass and gyroscope. You can hate it and be miserable or you can see it for what it is, another challenge in technology. Very few wizards can, or should, turn down the opportunities and the money that advancement provides them. It does bring issues, but you can never go back anyway. There is no noble path back down the mountain.

Being at the top brings its own set of problems. I recommend you view these problems as being only new technology parameters. In order to develop the technology, there are now other things to consider. You simply were not involved in them before.

The principles that brought you to the top will keep you at the top. Those include integrity, trust, honesty, quantification, checking and rechecking, evaluating, assessing, working and reworking, acknowledging wrong, eliminating affectations, effecting compromise, striving for perfection, making improvements, and recognizing all the faces of reality. Those assets cannot be bought. They come only from experience. Keep stepping.

Glossary of Terms

Abscissa	—	Horizontal axis on a graph, commonly denoted as the x-axis.
Alpha Test	—	A formal software test phase in which the simulated or actual product is tested by potential customers or an independent test team.
ATM	—	Asynchronous Transfer Mode, a telecommunications networking protocol.
Beta Test	—	A formal software test phase in which the product is tested by a limited number of users outside the development team.
BPSK	—	Binary Phase Shifting Key, a phase shifting scheme used to modulate data for transmissions.
Bullet	—	A typographical symbol, • , used to set off major elements in a discussion.
Charts	—	Common term for viewgraphs.
CDMA	—	Code Division Multiple Access, a channel access method used in radio communications.
CEO	—	Chief Executive Officer, highest ranking corporate officer in terms of management.
CFO	—	Chief Financial Officer, highest ranking corporate officer in terms of financial management.
COO	—	Chief Operations Officer, highest ranking corporate officer in charge of operations.
CPSK	—	Coherent Phase-Shift Keying, a technique used in digital transmissions.
CTO	—	Chief Technology Officer, highest ranking technologist in a company.
Decibel	—	Logarithmic measure of the pressure or power ratio as referenced to a standard.
Decision maker	—	Person with authority to approve or disapprove recommendations made by a technologist.
Density	—	In physics, the mass per unit volume. As used here, a reference to the amount of material on each viewgraph.
Differential equations	—	Mathematical equations involving differential calculus.
Flux	—	Amount per unit time. As used here, a measure of how much and how fast information is being transmitted.
Hardcopy	—	Printed version on paper of data held in a computer.

Hardware	—	Machines, wiring, and other physical components of an electronic system, especially as separate from the programming code (software) that drives the equipment.
IGFET	—	Insulated Gate Field Effect Transistor, a semiconductor device for amplifying or switching electronic signals.
IPVn	—	Internet Protocol Version n.
ISRO	—	Indian Space Research Organization, the space organization of India, roughly equivalent to the US NASA.
Lambda	—	Greek letter, λ, commonly used for the wavelength of a wave in the expression for the velocity.
Maxwell's Equations	—	Set of equations describing the electromagnetic force in nature and named after physicist James Clerk Maxwell.
MBA	—	Master of Business Administration, an advanced degree often obtained in executive programs.
MOSFET	—	Metal Oxide Semiconductor Field Effect Transistor, a semiconductor device for amplifying or switching electronic signals.
Nu	—	Greek letter, ν, commonly used to represent the frequency in a wave in the expression for the velocity.
Ordinate	—	Vertical axis on a graph, commonly denoted as the y-axis.
PHIGS	—	Programmer's Hierarchical Interactive Graphics System, a standard for rendering three-dimensional graphics.
PQFP	—	Plastic Quad Flat Pack, a type of integrated circuit packaging.
PowerPoint	—	Common software program used to produce viewgraphs for a presentation. A product of Microsoft Corporation.
R&D	—	Research and Development.
RPM	—	Revolutions Per Minute, a measure of angular velocity.
Softcopy	—	Legible version of data, especially as stored in a computer.
Software	—	Operating systems and programs used by a computer.
SATA	—	Serial Advanced Technology Attachment, a computer bus interface for connecting to mass storage devices.
SCMA	—	SubCarrier Multiple Access, a technique used in passive optical networks.
SCSA	—	Signal Computing Systems Architecture, an architectural framework for computer telephony integration.
SCSI	—	Small Computer Systems Interface, a set of protocols for physically transferring data between computers and peripheral devices.

Sea state	— a qualitative measure of the general condition of the surface of a body of water. A "zero" is calm and glassy, a "nine" is waves over 14 meters.
Slide	— a common term for a viewgraph.
Softcopy	— the presentation or other information in the bitwise form held by the computer.
Supersaturated	— a solution whose concentration of solute exceeds the solubility of the solvent. Any movement or other disturbance will cause almost instant crystallization.
TCP/IP	— Transmission Control Protocol/Internet Protocol, a descriptive framework for computer network protocols.
Technologist	— Scientist, engineer, or technician or any other individual whose primary interest and livelihood involves technology.
Viewgraphs	— Images, text, graphs, or any visual information that is projected on a screen for the broader audience to view. This is usually done simultaneously with the presentation.
Wizard	— Term used to describe any of a broad array of technologists. Meant to imply a person whose general thought processes are, in many ways, different from persons not involved in technology.

Index

Made in the USA
Charleston, SC
18 November 2011